T0281461

CABology: Value of Cloud, Analytics and Big Data Trio Wave

Nitin Upadhyay

CABology: Value of Cloud, Analytics and Big Data Trio Wave

 Springer

Nitin Upadhyay
Big Data Analytics and Information
 Technology
Goa Institute of Management
Goa, India

ISBN 978-981-13-4211-0   ISBN 978-981-10-8675-5   (eBook)
https://doi.org/10.1007/978-981-10-8675-5

Printed on acid-free paper

This Springer imprint is published by the registered company Springer Nature Singapore Pte Ltd.
The registered company address is: 152 Beach Road, #21-01/04 Gateway East, Singapore 189721,
Singapore

Dedicated
To
Avatar Meher Baba

*I have come to sow the seed of love in
your hearts, so that, in spite of all the
superficial diversity which your life in
illusion must experience and endure, the
feeling of Oneness, through Love, is
brought about amongst all nations, creeds,
sects and castes of the world*

Meher Baba

Preface

It was a 5'O clock in the evening in South Korea in the early winter of 2012. I was waiting outside the guest house in Seoul, to board a cab to meet my close friend—Kim. Right at 5:10, as prescribed by the cab service application, the cab arrived, and I boarded the cab. The interesting thing was that I could track the movement of the cab on the "*digital map*" without being bothered about calling the cab person to get the details about its location. It helped me not only to monitor the progress in the cab's arrival to the location but also to pick up the bag that I left in my room in the last minute before boarding the cab. The cab's driver welcomed me in a local language, which I could understand due to the language translator application support, and I reciprocated the same. With cold breeze outside, we started to move to the location, and in half an hour we reached the underway near to the destination. I was surprised to see that the display attached to the cab's dashboard showcased the underway "*digital map*" and prospective "*parking space*" and time to reach to the parking slot. Also, the display was also putting up the information about the CO_2 emission, health of the cab, performance, and rating of the driver. Technological developments, more specifically, a trio wave—cloud, analytics, and big data—are transforming the societal landscape the way we interact, collaborate, and communicate. Such a scenario is not unique but is scalable and can be found in varied dimensions of applicability and usage.

I entered the coffee shop and saw Kim waiting there. He was sitting at the corner table and was looking out the window. But no one could be seen outside the window. I said "Hello Kim" and repeated his name, but he didn't respond.

"The adoption of the technological elements became prevalent and utmost importance. We need to grow, create more meaningful services and products, and we need to churn the data to get hold of *what* is happening *when* to take actions and *where* we need to move?" said Kim. I could make out looking at his pale yellow face that something is not right.

Kim is an excellent technological leader and has shown his leadership in transforming the product industry during the Y2K problem and survived through the dot-com breakdown. But, now something is troubling him, and he is curious to understand the phenomenon of the elements of the business growth.

"Cost savings is not the only agenda, but exploration, experimentation, and development of new possibilities under the business drivers are the focus" murmured Kim.

We ordered Cappuccino, and he got a call on his mobile phone. He looked at the number and raised his eyebrow and said "Oh! Not again". He picked the call, and his conversation lasted for a quick 2 minutes, and then he ended the call. "I am having a board meeting tomorrow and need to discuss the technical and business issues". He sighed heavily. I could overhear his conversation that he is talking to a senior leader regarding the competitor's growth in achieving success by adopting facets of cloud, analytics, and big data.

"Why should I spend money to external entities to look after my IT service's needs, don't you think this is a core infrastructure that we need to own and be accountable for the business growth". He had many questions and looking for a silver bullet to see through the challenges and possibilities. He needs to clear the psychological boundaries. The old conventional life-cycle of "build vs. buy" is getting more imperative.

"Each department in the business organization is performing under the cut. I had conducted individual and group meetings. Individually people sounds to be motivated and are good at perfuming. Each one of them carries a work experience of more than five years. They all know the processes of the organization very well. But, somehow group thinking and working is not surfacing. That's why something needs to be done in that way. Moreover, change in the IT system towards the adoption of the trio wave is giving more challenges for the organization. People are not prepared for this change. I felt the resistance in one of the meetings where an IT manager said:

"This is ridiculous, we know all business needs and requirements. How can we trust the third party vendor for storing our data and processes? What value we can get?"

Kim took a long pause and asked, "What do you think, aren't theses concern genuine? How we can understand the whole ecosystem?" Further, he said, "it's getting more complicated, could you through some light on this aspect?" Well, Napkin tray comes handy which was lying on the corner of the table, and I picked few of them to put my ideas about the aspect and drew few pictures. I told him, "the time has gone where one needs to take corrective action. You, cannot wait to see the losses, setbacks, low stock market price. You need to understand, plan for the future, by knowing your status-quo and where you want to be in the business context. Don't worry we will figure out and charter the path and roadmap to reach there at the destination."

We ordered another cup of cappuccino and some snacks to satisfy our urge of hunger. So, what do you think? Will I be able to deal with this new agenda of transforming the whole IT ecosystem? Eagerly enquired Kim.

I worked with Kim for the next two months to help him to transform his business by the adoption of cloud, analytics, and big data. There are many alike Kim who would like to initiate their journey into the CAB space but are hesitant as the proper guidelines, directions, framework, roadmap, and strategies are not available handy.

This book elaborates the various facets of my experience of working with many organizations and strategizing their businesses to embrace the cloud, analytics and big data for generating business value and achieving transformation.

Why this book

Cloud, analytics, and big data (CAB) have transformed the business. It encompasses an exciting new set of tools, techniques, frameworks, and platforms for modern, data-powered applications to generate business value. A fundamental shift needs to occur in thinking about how we manage, control, compute, and analyze data to generate the business value. This book will prepare stakeholders of the digital era for that shift in thinking by providing an elaborative discussion on cloud, analytics, and big data in a readable, straightforward fashion. Most of the books written on cloud, analytics, and big data treat the topics in silos. Also, the books are dominated by few select experts and wizards who tend to present the subject either into too detail or too complicated. Thus, making it for the readers unnecessarily over technical and intimidating.

The cloud, analytics, and big data is no longer a "cup of tea" for technical savvy rather its usage and implication cut through managerial, decision making, economics, and technical core in theory and practice.

The book—"CABology: Value of Cloud, Analytics and Big Data Trio Wave"— presents an optimal balance of practicing the set of strategies, techniques, methodologies to derive and communicate valuable and actionable performance measures, benefits and insights. CABology is to [study] focus on the art and science of optimizing the business goals of delivering true value and benefits to the customer through the cloud, analytics, and big data.

A plethora of digital content is available on the Internet, which will take more than a lifetime for a reader to go through and grasp the concepts. Also, extensive print literature is available which is either too specific targeting only a small group or too general to focus on a broader group, limits the benefit of learning, experimenting and exploring concepts, tools, and techniques.

This book considers the apt and balanced way of addressing the topics in a lucid and readable fashion.

In this book, I introduced a broad array of concepts and approaches that one can use in many ways, such as to

- Explore the status of CAB's evolution and now
- Understand the utility of CAB ecosystem
- Identify and set the true value and benefits for the customers
- Assess and evaluate CAB tools, techniques, and methodologies
- Develop CAB strategies to align business goals
- Set the CAB proposition for optimal benefits (win-win)

I have organized this book in a structured, modular, and easy-to-read access format. For the readers, it provides benefits as a potential guide and a reference for critical pointers.

Chapter 1: A triology—Evolution and Now

We are rapidly unfolding and discovering the future. This chapter proposes to discuss the evolution and now the status of CAB covering—its challenges, limitations, usage, potential, and value. The chapter will provide relevant scenarios and examples.

Chapter 2: CABology—What is in it?

One of the most critical and strategic resources that organization is looking at is to develop and find timely, valuable, and insightful actionable decisions. This chapter proposes to introduce CABology, which means to [study] focus on the art and science of optimizing the business goals in delivering true value and benefits to the customer through the cloud, analytics, and big data. The chapter will provide relevant case studies.

Chapter 3: CAB Framework—The Fabric

Organizations no longer need the traditional IT services in the changing world of digital business and fierce competitions. They require that their IT infrastructure, whether they are the consumer or producer of the resources, should enable them to be agile, more powerful, and cost-effective. Today's business ecosystem requires that the IT infrastructure must support the change in current and future business needs and strategies. The chapter will discuss the CAB framework principle and concept in length and provides relevant case studies. The CAB frame-work (fabric) helps the organization to operationalize the value of big data.

Chapter 4: CABonomics—The True Value

In this chapter, you will find the discussion on the economics of CAB for sustainable and competitive advantage. The mere availability of a set of resources, tools, techniques, frameworks, and methodologies does not provide the true value and actionable insights. One must also understand and learn the various strategies to economize.

Chapter 5: CABevaluation—What is Right for Me?

The world has put across us various components of the CAB under propriety, open source and otherwise. It is challenging for an organization to understand and leverage the maximum benefit out of CAB components. In this chapter, you will explore in detail about the plethora of CAB components alternatives, evaluation criteria, and strategy to adopt/acquire/integrate into the business environment.

Chapter 6: CABfication—The Explanation

One must operationalize on CAB framework by defining the key and core business operations. In this chapter, the focus is to identify the key and core operations of the business that could rely on a CAB for generating business value. A 3M's model—

measure, manage, and monetize—is proposed that will help the stakeholders to drive the monetization activity.

Chapter 7: CAB Proposition—The Way Forward

This chapter will discuss the various propositions for consumers and producers of the CAB (in part or full) which will help them to emerge as a leader. It also creates a pathway for any organization to build a scalable and sustainable business. The CAB drivers are necessary to drive the CAB adoption and sourcing to gener-ate the business value. Readers will explore the critical components of the CAB drivers.

Chapter 8: CAB Implications—The Affairs

Selecting and implementing CAB culture and infrastructure in the organization, as an option is less straightforward. In this chapter, the CAB maturity model (CABMM) is proposed that will help the organization to understand status quo, and by utilizing it, an organization can move incrementally or leapfrog to attain sig-nificant competitiveness.

Chapter 9: CAB Control—The Power

In the digital world, the organization is getting data from varied sources, working with the orchestration of multilayers technology in various jurisdictions. It imposes an organization to have proper and effective CAB control to reassure so that that the CAB[ology] framework used by an organization operates as intended, that data and analysis are reliable and that the organization utilizes the controls mechanisms and governance structure to drive the CAB adoption or sourcing.

Chapter 10: CAB success stories

In this chapter, select few case studies have been discussed. Readers will be able to understand the challenges, opportunities, and useful lessons of CAB adoption or sourcing.

Appendix: Worksheet activities

In this section, various worksheets are proposed to help the CXOs, leaders, con-sultants, and senior executives to explore and understand the pathway needed to transform the organization. It will help the organization entering into the CAB space to focus on guiding principles to execute its efforts and investments as per the corporate priorities considering culture, governance and transition, monetization and value generation, and to mitigate any risks.

Goa, India   Dr. Nitin Upadhyay
June 2018

Acknowledgements

To my wife, Dr. Shalini Upadhyay, I owe my indebtedness for always supporting me in my ups and downs. Her never-ending and unconditional love cannot be expressed in words. I am thankful to her for teaching me the virtues of perseverance, dedication, and determination.

I am beholden to my sweet little angel and dearest daughter, Ms. Meherika Upadhyay, for making my life so beautiful with her smile, laughter, and innocence. For every night asking me "will you study and write chapters for your book?", which showed that even in this tender age she understood the significance of my work and that she was concerned about me. Not only am I indebted to her but also apologetic for stealing her time in order to realize this work. A special thanks to my dog "Brownie" for spending endless hours with me during the project and showing love and care and bringing energy.

I am thankful to all the reviewers who have provided necessary and appropriate comments for making this book as apt as possible. I acknowledge my students, fellow colleagues, peers, and researchers for their healthy discussions, suggestions, and necessary cooperation toward the completion of this work.

I am thankful to the Director, Goa Institute of Management, Goa, for providing me the conducive environment necessary to accomplish such a project.

Quite importantly, I would like to express my deep gratitude toward Springer Nature, publisher, for providing me with an opportunity to write a book on such an essential topic, and for monitoring the progress and providing valuable suggestions for the improvement of the chapters from time to time.

Finally, I express my gratitude to all those who have directly and indirectly helped me in this project.

Dr. Nitin Upadhyay

Contents

About the Author

Dr. Nitin Upadhyay is a researcher, inventor, innovator, consultant, leader, coach, academician, and a prolific writer. Over the years, he has engaged with select top Fortune 500 companies. He is a leading authority and speaker on innovation, design, cloud computing, big data and analytics, future technology, and user experience. He is currently an associate professor in the area of information technology and is a core member of the Big Data Analytics programme, Goa Institute of Management, India. He is also the chair and head of the Centre for Innovation at Goa Institute of Management. He has wide industry, academic, consultancy, and research experience and is an executive member and chair of cloud service-level agreements (SLAs) for the Cloud Computing Innovation Council of India. He has contributed to numerous peer-reviewed publications/presentations/posters and talks and eight books.

He has received many awards and recognition nationally and internationally. He is listed in the *Who's Who in the World, Who's Who in the Asia, Top Innovators of the World (IBC Cambridge), and received multiple awards such as—Research Shepherd, Star Researcher and Innovator, Outstanding Scientist, Lifetime achievement award USA, Best Research (Intl. Conf. South Korea).*

He also paints on canvas (acrylic/oil) and has exhibited his paintings to the wider audience. On the outset, he participates in marathon, plays guitar, contributes to literary corpus, and is also a voracious reader.

Chapter 1
A Triology—Evolution and Now

The world doesn't change in front of your eyes, it changes behind your back.

—Terry Hayes, I Am Pilgrim

Thus, from the war of nature, from famine and death, the most exalted object which we are capable of conceiving, namely, the production of the higher animals, directly follows. There is grandeur in this view of life, with its several powers, having been originally breathed into a few forms or into one; and that, whilst this planet has gone cycling on according to the fixed law of gravity, from so simple a beginning endless forms most beautiful and most wonderful have been, and are being, evolved.

—Charles Darwin, The Origin of Species

Introduction

The digital era has developed a radical competition among the businesses (Hockey, 2007). Each business would like to understand more about its customers, channels, and engagements and would want to leverage the digital space to succeed. Triology —cloud, analytics, and big data are creating a tectonic shift in the businesses and society. The only option is to embrace it, grow with it, and get transformed or to reject it and left behind.

Amazon, a business giant, has created a swift in reshaping the businesses those are performed in the digital space. It offers a plethora of services that vary from B2B, B2C, and C2C (Bende & Shedge, 2016). Amazon has redefined the "book" industry in a "datafied" way. It could manage to turn each book into a tabulated fashion and perform metric-based operations to achieve effective analytics. The Amazon's "Kindle" service has cornered the book industry. More and more publishers are publishing Kindle ready book to be positioned into a flourishing and promising digital sector thereby be present close to the readers' space. Amazon Kindle has access to massive amount of data of its every reader. The way that each Kindle reader is engaged with the Kindle service is astonished. Some of the

© Springer Nature Singapore Pte Ltd. 2018
N. Upadhyay, *CABology: Value of Cloud, Analytics and Big Data Trio Wave*,
https://doi.org/10.1007/978-981-10-8675-5_1

questions are now well explained by the Amazon, for example how much time a reader is taking to complete the book? How many words per minute being read by the reader? What is the reading taste of the reader? What is probable books reader interested to learn? And many more. The whole lot of data provide Amazon to develop multiple characteristics to microtarget readers and then provide them with the customize recommendation or type of books to read.

Travel agencies are also embracing digital space to thrive their businesses. Online travel agents (OTAs), Expedia, Priceline, Airbnb, and now Google, all are now bundling packages and exploiting customer's data to provide them the customized choice of services (Baggio, Sigala, Inversini, Pesonen, & Eds., 2013; Holmberg & Cummings, 2009). The most astonishing experience for the fulfillment of the customer's expectation journey starts from the [intention of the] data collection stage. The process of knowing customers is getting better as now the travel industries get hold of far more richer data about customer profiles, geolocations, choices, and user experience. For example, it is now easy to figure out at what stage the process resulted in the termination, whether customer just walked away? Or looking at something else? Or did they just went through until payment and exited the process? All such rich data are now helping these travel industries to dig deeper and get the insights about customer travel booking experience, and for achieving the operational excellence.

The large OTAs have a significant advantage over the data access and its utilization than the airline industry, as customers approaching them shares the massive amount of data. For example, for an airline industry the booking is just for a source to destination, but for an OTA a customer is worth more. An OTA customer might be interested in Flight + Hotel or Flight + Hotel + Car rental services. Thus, the OTAs can understand and classify customers who all are interested in single, multiple, or package options. Once the customer data is available, the whole process can be customized, tailored-made to microsegment the customers. The organization focuses on creating data factory to understand customer and to provide customized services and offerings. A data factory related to the airline industry is shown in Fig. 1.1.

People leave long trails of a digital footprint when they travel. For example, data related to their GPS coordinates during their movements in the physical space, interlay calendar details for the travel and meetings, and online purchase details is available (Balakrishna, 2012).

It is important to know that customer data is fragmented, and one who holds this fragmented data can leverage it to leap generating the business value. A fragmented customer data across numerous different systems can be incomplete or inaccurate and also be proprietary or off the shelf.

In a simple data-away brick-and-mortar business model, it was challenging for the retailers to know the customers as they were nameless, faceless, identity-less, and uncategorized entities (Liu, 2013). Retailers nowadays can understand customer lifetime values and use customer's data to make it more useful. The retailers with the help of insights are then able to treat each customer differently and offer

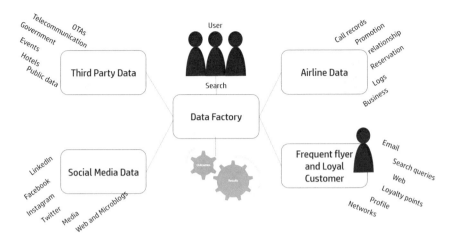

Fig. 1.1 Data factory

services appropriately based on customer choices, preferences, perceived behaviors, and many other relevant attributes.

In North America, one of the retail insights company, Brickstream, is pioneered in understanding customer physical movements and expressions in a physical retail space. The technology by Brickstream is simple and powerful that can capture data related to the people as they enter, move through the retail space, and exit the retail environment. Brickstream can seamlessly capture, collect, integrate, and analyze data by using edge analytics and cloud technology. Some of the company's offering includes:

- Path tracking analytics embedded in the Brickstream smart devices and BehaviorIQ, offering rich people metrics across a series of linked devices.
- BehaviorIQ, an extendable software platform for data management and integration with the enterprise data warehouse.
- Real-time dashboards, reports, alerts, and analytics tools.

Retail Next provides in-store analytics to deliver actionable insights to the retailers, stores to engage customers effectively. The RetailNext technology uses multiplatform technology by integrating best-in-class video analytics, wireless detection of mobile devices, multistore data sources, and even weather data to analyze factors impacting customers shopping behaviors.

Equifax once a credit card company now has learned to harness the data of more than 800 million consumers to take an informed decision. The company now supports multiple industries and provides valuable services to the consumers. Healthcare technology and consulting company, Geneia, helps its customers to provide better and economic healthcare services. Geneia could manage to gain excellent user experience on its service offerings through its software Theon. Theon collects, integrates, analyzes, and synthesizes data for care team members and patients and help them to take an informed decision on early but subtle signs of ailments, pains, disease, or healthcare needs. Theon has broad coverage of the stakeholders such as—C level executives, health managers, practitioners, nursing coordinators, chemists, insurers to enable them to consider more informed decisions, choices, and actions.

> *If you have a basketball and a baseball 14 feet apart, where the baseball represents the moon and the basketball represents the Earth, and you take a piece of paper sideways, the thinness of the paper would be the corridor you have to hit when you come back.*
>
> —David Scott
>
> Astronaut to the Gemini 8, Apollo 9, and Apollo 15 Missions.

Triology—Cloud, Analytics, and Big Data

Cloud, analytics, and big data (CAB) do not exist in silos, and the combination brings out the value—big intelligence, Fig. 1.2. The Triology considers various management and operational aspects of the cloud, analytics, and big data. It also deals with the incorporation of CAB strategy document in aligning with the organization's vision. It is best explained by the Google's Director of Research, Peter Norvig, by saying that "We don't have better algorithms. We just have more data." Nonetheless, we also have the more efficient infrastructure to work with algorithms (intelligence processes) and (big) data. Otherwise, it is not possible to generate just-in-time or (near) real-time analytics. In this section, we will explore the evolution and challenges associated with cloud, analytics, and big data.

> As the wonder and mystery of data science begins to wear off…so will the hype (and high prices!) around the data science tools.
>
> —Rupinder Goel, GLOBAL CIO, TATA

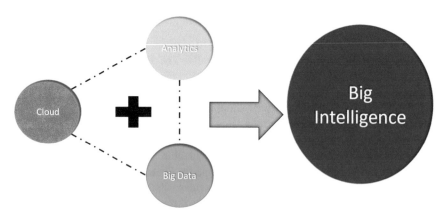

Fig. 1.2 Big intelligence

Evolution of Data

One of the oldest known recorded data is almost 30 thousand years ago in central Europe. Let us for the simplicity of the calculation, we assume that an average person lifespan is around 100 years (I am being more generous). Then it is almost before 320+ generations, the first data record keeping happened. Someone scratched wolf bone with almost 57 notches on it. No one knows whether it was done by a same person or different or what is the purpose of it or what it denotes. Nonetheless it is the breakthrough the way the data record was materialized. Now imagine you wish to count wolf in four regions and their movement on the daily basis. How would you know that every next day morning all of the wolf are in the same region and not moved out, until you identify and count them? A solution is to notch for each wolf and thus maintain a count. This count can also be used to record killing count of a wolf or losing of a wolf from that region.

More surprising, dated back 32000 years in the Ice Age Cave, now known as Czech Republica, initiated the breakthrough idea of digital data. Now, the data about wolf, supported by the International Wolf Centre (IWC) in Minnesota USA, can be downloaded from the Internet-enabled devices. The IWC with the help of Internet of Things (IoT) enabled collar could track movement, patterns, and many more.

Big Data

A mobile that we carry has a far superior computing and storage power as compared to the computers that existed in the world five decades ago. Interestingly, the NASA Apollo era's ingenious computer has guided astronauts across 356,000 km of space from the Earth to the Moon. The system helped them to return safely. Don't you think that with today's infrastructure we can run hundreds of such projects?

Data Challenges

The organization should mandate, regulate, and validate the quality of the data and (intelligent) processes used to work with the data. Having an extensive data does not mean that an organization has an advantage over its competitor (Demchenko, De Laat, & Membrey, 2014). Similarly, having a data lab that is fully equipped with Hadoop and Spark ecosystem does not give a benefit to the organization until challenges associated with the data are resolved. Certain problems are highlighted below to focus on before taking any action, Fig. 1.3.

Data Vision

Key business holders must envision the 4W's and 1H data vision framework of data. Figure 1.4, which should be plugged into enterprise-wide system to provide benefits to the associated stakeholders. The challenges are many as one needs to decide on the priority, impact, and need of the 4W + 1H data vision framework.

Data quality

Quality of data is critical to provide the required results and the outcomes. Few characteristics of poor data quality are listed below:

Noise

Noisy data is mostly influenced by and contain misleading and conflicting information.

Dirty

Data is treated as dirty when it contains many missing values, fuzzy information, factors having a large depth of levels.

Sparse

At this level, data captures mostly zero values and very few actual values. This type of data is scattered and needs specific techniques to perform operations.

Inadequate

It refers to the data that is either incomplete or insufficient to perform any useful operations.

Data can be preprocessed before the data is ingested into a storage area or kept at the staging area for further processing.

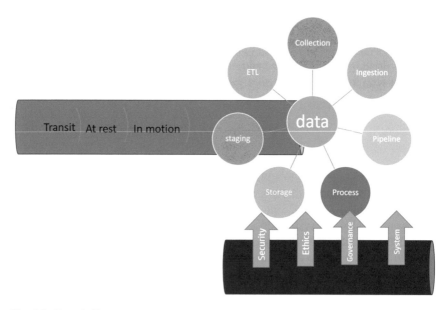

Fig. 1.3 Data challenges

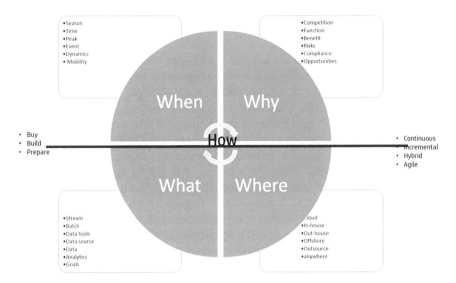

Fig. 1.4 Data vision

Data Security and Governance

Data security and governance is paramount necessary for some of the industries like banking, health care, insurance, and government (Greengard, 2014; Morabito, 2015). It has been reported that Cambridge Analytica has harnessed unethically more than 50 million Facebook subscribers' data to target them with the personalized political campaigns and advertisement resulting in the fabrication of the US election (Liptak, 2018).

Data security does not tell the full story of the data governance instead it captures only the part of it. One of the critical challenges to generate the business value with data is to answer the business questions. But, how many business questions are preconceived before the acquisition of the data and how many keep adding as a project matures. In most of the cases the business questions, and the needs leading to the desired results often change during the [big data] project life cycle which provides quite a challenge to data exploration.

> Once the twin challenges of generating high-quality insights and effectively delivering these insights have been resolved, the acceptance and adoption of Analytics will automatically improve, especially when there is sponsorship of Analytics initiatives at the highest levels of the organization.
>
> —Vijay Kumar, Former Analytics Head, SBI

Evolution of Analytics

As we entered the nineteenth century, the "clockwork universe theory" propounded the understanding of the universe by comparing universe to a mechanical clock. Further, the gears of the clock are governed by the laws of physics and contribute to the prediction of the aspects related to the machine. In the early thirteenth century, a similar concept was discussed by Sarcobosco's medieval text that describes the world as "machine Mundi" means the machine of the world. In 1820, Pierre Simon Laplace proposed that there exist certain patterns for the errors associated with the scientific observation (Rao & Jain, 2016). Nonetheless, this simple thought triggered the statistical evolution. The deterministic worldview got affected by the precise measurement, as instead of achieving the diminishing error with the preciseness of the measurement the error grew challenging the deterministic view. It leads to the disruption of the deterministic approach with the probabilistic approach proposed by Karl Pearson. He identified that the measurements, having inherent randomness, are probabilistic, while the errors associated with the measurements are not probabilistic. Pearson elaborated that the "ideas" of science, such as mathematical laws that describe the laws of motion, are not observable and hence not precise, but can be estimated by using formulas and data. Perhaps the foundation was laid, and a transformation happened from the deterministic to the probabilistic perspective.

In the early twentieth century, Sir R. A. Fisher, often refer to the "Father of Statistics," published remarkable papers and performed studies to elaborate on the concept to conduct experimental design by factoring out the external effects that are not considered to be an inclusive part of the experimental design. The "significance" testing was contributed mainly by him by which one can test the significance of the factors. In 1933, Jerzy Neyman and Egon Pearson extended the significance testing by formulating the hypothesis testing and structuring the decision-making process. The contributions of John Tukey in the inference theory and interpretation made the data analysis to appear as a mainstream discipline. In the latter half of the twentieth century, the development of mainframe, personal computers, handheld devices, and Internet revolutionized the way the data is captured, stored, and processed. Scientists, business people, and academia came together to understand and identify the value of the data. It merely leads to the development of more powerful, flexible computing and storage architectures, different techniques to capture, store, process, and visualize data, and the deployment of models to solve complex use case scenarios.

Analytics

Analytics is simply the use of data, statistical and quantitative analysis, intelligent models, evidence, and fact-based decision making at scale and speed to drive and deliver actionable decision and insights. The author, Davenport and Harris, propounded four different eras of analytics—Analytics 1.0, Analytics 2.0, Analytics 3.0, and Analytics 4.0, as part of the analytics evolution (Davenport & Harris, 2007), Table 1.1 and Fig. 1.5.

Analytics is also classified by the type of (business) context and the problem it attempts to solve (Fan, Lau, & Zhao, 2015; Hasan, Shamsuddin, & Lopes, 2014; Singh & Singh, 2012).

- Descriptive: It deals with the descriptive attributes of the context and attempt to solve questions like—What happened, how many, and how often. For example, what's the benefit and potential of leveraging descriptive analytics to understand the business operations, staffing, sourcing, materials management, market campaigning
- Diagnostic: It deals with the diagnostic attributes of the context and attempt to solve questions like—where exactly is the problem, what is the cause, what actions are needed. For example, what's the benefit and potential of leveraging diagnostic analytics to diagnose the business operations, staffing, sourcing, materials management, market campaigning
- Predictive: It deals with the predictive attributes of the context. It attempts to solve questions like—what will happen. For example, what's the benefit and potential of leveraging predictive analytics to predict the business operations, staffing, sourcing, materials management, market campaigning
- Preventive: It deals with the preventive attributes of the context. It attempts to solve questions like—what should be avoided if the trends continue? For example, what's the benefit and potential of leveraging predictive analytics to prevent the service and product failures, fraud, crime, physical, and monetary loss?
- Prescriptive: It deals with the prescriptive attributes of the context. It attempts to solve questions like—what is the best can happen? For example, what's the benefit and potential of leveraging predictive and prescriptive analytics to prescribe and optimize the business operations, staffing, sourcing, materials management, market campaigning?
- Autonomous: It deals with the autonomous attributes of the context. It attempts to build models having minimal human intervention and are self-serviceable, cognitive, and driven by machine learning. Try to answer, what's the benefit and potential of leveraging autonomous analytics to recommend the customers for buying, upgrading, changing, modifying service and products?

Table 1.1 Analytics era

Analytics x.0	Description
Analytics 1.0	Largely focus on descriptive analytics. Primary business context to understand what happened in the past. Usage of reports and visuals
Analytics 2.0	Incorporating big data in the decisions. The focus shift on the technology and business landscape. The inclusion of prediction for decision making
Analytics 3.0	Big and small data being adopted as mainstream for the analytics. Utilization of prediction outcomes in prescriptive and preventive analytics and modeling. Execution of actionable decisions and insights
Analytics 4.0	Self-serviceable, cognitive, and machine learning-driven analytic models. Minimal human intervention is required

Fig. 1.5 Analytics era

Big Data Analytics

The big data era is witnessing the convergence of machine, people, and organization data to stimulate and develop actionable and informed decisions. Figure 1.6 deals with the six critical characteristics of big data that is impacting and contributing to analytics. The volume refers to the scale of the data of that is available, accessible, and utilized for the analytics. The variety deals with the type of data—structured, semi-structured, and unstructured. The proliferation of social media applications has contributed to the growth of unstructured data—videos, audios, text, and images. The streaming and edge processing became relevant due to the requirement of mission and critical projects, such as fraud detection, and it represents the velocity characteristics of the big data. The veracity of the data refers to the property of the data covering biases, noise, and abnormality. Valence refers to the complexity of the data when induced by relationships and connectedness, such as a social network. Finally, the value of the data is the treasure that every organization is trying to achieve.

Batch Analytics

In batch analytics, the organization is interested in inducing and ingesting the bulk of data and processing it in a timely schedule manner to generate the output. The focus at this level is twofold. Firstly, to identify the characteristics and property of the batch to be considered for the processing and secondly, the business use case where the batch can be utilized for generating the business value. For example, customer batch profile data can be used for preparing ad campaigns, putting offers, and strategizing service delivery channels.

Streaming Analytics

Mission critical and high risks tasks do not have a luxury of time and thus needs to be monitored, analyzed, managed, and controlled on a real-time or near real-time basis. However, (near-) real-time analytics looks promising to perform better and reduce risks but also opens new security threats and has financial implications. Furthermore, challenges are more when data is analyzed based on location. For example,

   Edge Analytics—It deals with the processing of the data that is collected within the device (gateways, sensors, machines, etc.)

   In-stream/in-motion analytics: It refers to the pipeline analytics where the data being analyzed between the device and the server (i.e., network security logs)

   At-rest analytics: It is separate from batch analytics in a way that it analyzes the data when the data reaches at the server for processing (e.g., consumer real-time event streaming)

Fig. 1.6 Big data characteristics

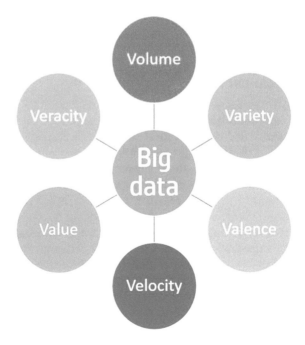

Challenges

The analytics vary depending upon the context of usage and application/domain industry (Grossman, 2018; Singh & Singh, 2012). For example, in a healthcare industry, some (near) real-time analysis is performed for checking of blood sugar level, BP, heart rate while in automotive industry sensors and onboard electronic components are used to send data and perform analytics of automobile's speed, oil pressure, brake and engine health, etc. Perhaps most of the online service company utilizes customer reviews to perform opinion and sentiment analytics.

More challenges such as data model and decisions, interpretation, documentation, and justification often appear in the regulation industry where advanced analytics is performed on the data. Another problem on the analytics spectrum is on the deployment of the model in the production environment. It has twofold implications. Firstly, on the use of actual streaming data into the production environment which causes induction of noise when more data is included by resulting in deterrent performance, and secondly, the transformation of the environment from the personal development to the production environment.

> Line-of-business leaders everywhere are bypassing IT departments to get applications from the cloud (also known as Software-as-a-Service, or SaaS) and paying for them like they would a magazine subscription. And when the service is no longer required, they can cancel that subscription with no equipment left unused in the corner
>
> —Daryl Plummer, Managing Vice President and Distinguished Analyst at Gartner

Evolution of Cloud

Cloud computing did not appear on any single day but its governing principles, context, and model evolved since the development of the first machine to store, process, and output the data. In the mainframe era, multiple users could access the central resource. Further, when IBM in 1970 came with the technology of the virtualization, it was possible for the machine to host multiple operations systems at the same time and offer a viable choice to its end users to plug and use any one of them. Virtualization became a key driver to revolutionize communication and information infrastructure. In early 1990s, telecommunication industries adopted the virtualization concept and started offering private virtual network connections that enabled companies to provide shared connections to its clients instead of single dedicated point-to-point connections. Further, concepts like grid computing to solve large problems with parallel computing; utility computing to offer computing resources as metered; Software-as-a-Service to provide network-based subscription to the applications contributed to the cloud computing model.

Cloud

The cloud computing refers to the anytime, anywhere, on-demand, metered access to IT resource delivered dynamically as a service (Vouk, 2008; Weinhardt et al., 2009). Cloud computing has typically four deployment models and offered services of four types.

Cloud Deployment Models

It deals the way that the cloud service is delivered to the user.

- Private: It refers to the cloud that is available exclusively to use by the single organization. The cloud can be controlled, managed, and provisioned by the organization, third party, or a combination of them. The cloud can exist within or outside the organization premise.
- Public: The cloud is owned, managed, maintained, controlled, and operated by academia, organization, business, or combination of them, which is provisioned for open use to the general public.
- Hybrid: It refers to the cloud that is the combination of more distinct cloud infrastructure—private, public, and community to provide economic competency to the organization.
- Community: The cloud is available to use for the specific community having similar concerns. It may be owned, managed, maintained, controlled, and operated by one or more organization of the community, third party—academia, organization, business, or combination of them, which is provisioned for open use to the specific community.

Cloud Service Models

It denotes the way that the cloud services are available for use (Tao, Zhang, Venkatesh, Luo, & Cheng, 2011; Upadhyay, 2016):

- Software-as-a-Service (SaaS): It refers to the software application that is directly available to use by the user through the Web/Internet access. For example, Google's Gmail application is a SaaS and used widely by the users. In the mobility dimension world, "SaaS" is likely to be considered "an app," as the user interacts with the user interface through the mobile device and rest of the process and operation happens on the cloud.
- Platform-as-a-Service (PaaS): It provides environmental support to configure, deploy, and manage applications. In some cases, companies have adopted PaaS platform to give an extension to their software applications so that application developers and customers to create required customized applications and services. For example, Force.com is the extension and complements the Salesforce.com.

- Infrastructure-as-a-Service (IaaS): It offers complete control to configure, manage, and deploy the infrastructure services such as load balancing, scaling, content delivery network control, VMs.
- Anything-as-a-Service (XaaS): These are the optional service model that sits on top of the service stack. It targets special services like Data-as-a-Service, Business-as-a-Service, Data analytics-as-a-Service, and many more.

Challenges

Managing the various aspects of the cloud infrastructure and services becomes a challenging job for many organizations. It is so, as many of the issues are hidden or not disclosed to the customer. Some infrastructure and service challenges include storage, computation, governance, and service-level agreements. Let's explore each of these challenges.

- Storage: Every organization is running in the race to capture massive data to perform useful computations. It is quite challenging to provide orchestrated storage solution considering multiple architectures, tools, and storage platform solution in the marketplace.
- Computation: Once you have all data, then you would like to operate that so that you can get some meaningful insights and decision action points. Most of the analytical tasks are computationally intensive. The resource requirement is dynamic as per the need of the analytic requirement.
- Governance: It is vital to the success of organization's cloud adoption or sourcing strategy. It is challenging to avoid fragmented implementation impact of the cloud services; organization needs to develop the business value plan by considering the external forces, current infrastructure, resources competency and capability, and culture.
- Service-level agreements: Management and execution of service-level agreements are critical for the cloud adoption. Issues like vendor lock-in, service outage, data security, privacy, and safety are challenging and sometimes difficult to address. Each jurisdiction has different rules for holding, transmitting, and processing the data. Also, some countries promote and put constraints on building the CAB solution by the country-specific citizens only.

Concluding Remarks

The Triology—cloud, analytics, and big data have a high potential for identifying, generating, and driving the business value. Organization needs to understand the benefits, challenges, opportunities, and risks associated with the Triology. Big intelligence will play a critical role in guiding organizations to consider actionable and insightful decision and actions.

References

Baggio, R., Sigala, M., Inversini, A., & Pesonen, J. (Eds.). (2013). Information and communication technologies in Tourism 2014. *EProceedings of the ENTER 2014 Ph.D. Workshop*, 1–146. https://doi.org/10.1007/978-3-319-03973-2.

Balakrishna, C. (2012). Enabling technologies for smart city services and applications. In *Proceedings—6th International Conference on Next Generation Mobile Applications, Services, and Technologies, NGMAST 2012* (pp. 223–227). https://doi.org/10.1109/NGMAST.2012.51.

Bende, S., & Shedge, R. (2016). Dealing with small files problem in fadoop distributed file system. *Procedia Computer Science, 79,* 1001–1012. https://doi.org/10.1016/j.procs.2016.03.127.

Davenport, T., & Harris, J. (2007). *Competing on analytics: The new science of winning*. Harvard Business School Press.

Demchenko, Y., De Laat, C., & Membrey, P. (2014). Defining architecture components of the big data ecosystem. In *2014 International Conference on Collaboration Technologies and Systems, CTS 2014* (pp. 104–112). IEEE Computer Society.

Fan, S., Lau, R. Y. K., & Zhao, J. L. (2015). Demystifying big data analytics for business intelligence through the lens of marketing mix. *Big Data Research, 2*(1), 28–32. https://doi.org/10.1016/j.bdr.2015.02.006.

Greengard, S. (2014). Big Data = Big Challenges. *CIO Insight*, 1.

Grossman, R. L. (2018). A framework for evaluating the analytic maturity of an organization. *International Journal of Information Management, 38*(1), 45–51. https://doi.org/10.1016/j.ijinfomgt.2017.08.005.

Hasan, S., Shamsuddin, S. M., & Lopes, N. (2014). Machine learning big data framework and analytics for big data problems. *International Journal of Advances in Soft Computing and Its Applications*, 6(2).

Hockey, S. (2007). The history of humanities computing. In *A companion to digital humanities* (pp. 1–19). John Wiley and Sons.

Holmberg, S. R., & Cummings, J. L. (2009). Building successful strategic alliances: Strategic process and analytical tool for selecting partner industries and firms. *Long Range Planning, 42* (2), 164–193. https://doi.org/10.1016/j.lrp.2009.01.004.

Liptak, A. (2018). Cambridge Analytica's use of Facebook data was a 'grossly unethical experiment'. https://www.theverge.com/2018/3/18/17134270/cambridge-analyticas-facebook-data-underscores-critical-flaw-american-electorate.

Liu, A. (2013). The meaning of the digital humanities. *Pmla, 128*(2), 409–423.

Morabito, V. (2015). Big data and analytics: Strategic and organizational impacts. *Big Data and Analytics: Strategic and Organizational Impacts*, 1–183. https://doi.org/10.1007/978-3-319-10665-6.

Rao, H., & Jain, D. (2016). Perspective on fast-developing field and its expected impact on science now and in the future. *Informs, 40*(6).

Singh, S., & Singh, N. (2012). Big data analytics. In *2012 International Conference on Communication, Information {&} Computing Technology (ICCICT)* (pp. 1–4). https://doi.org/10.1109/ICCICT.2012.6398180.

Tao, F., Zhang, L., Venkatesh, V. C., Luo, Y., & Cheng, Y. (2011). Cloud manufacturing: A computing and service-oriented manufacturing model. In *Proceedings of the Institution of Mechanical Engineers, Part B: Journal of Engineering Manufacture 225*, (pp. 1969–1976). https://doi.org/10.1177/0954405411405575.

Upadhyay, N. (2016). SDMF: Systematic decision-making framework for evaluation of software architecture. In *Procedia Computer Science* (Vol. 91). https://doi.org/10.1016/j.procs.2016.07.151.

Vouk, M. A. (2008). Cloud computing—Issues, research and implementations. In *Proceedings of the International Conference on Information Technology Interfaces, ITI* (pp. 31–40). https://doi.org/10.1109/ITI.2008.4588381.

Weinhardt, P. D. C., Anandasivam, D.-I.-W. A., Blau, D. B., Borissov, D.-I. N., Meinl, D.-M. T., Michalk, D.-I.-W. W., et al. (2009). Cloud computing—A classification, business models, and research directions. *Business & Information Systems Engineering, 1*(5), 391–399. https://doi.org/10.1007/s12599-009-0071-2.

Web References

John of Sacrbosco. (1974). *On the Sphere, quoted in Edward Grant, A Source Book in Medieval Science* (p. 465). Cambridge: Harvard Univ. Pr.

https://en.wikipedia.org/wiki/Clockwork_universe

Harsha Rao and Deepali. Historical perspective on fast-developing field and its expected impact on science now and in the future. Jainhttps://www.informs.org/ORMS-Today/Public-Articles/December-Volume-40-Number-6/The-evolution-of-analytics

https://www.ibm.com/blogs/cloud-computing/2014/03/18/a-brief-history-of-cloud-computing-3/

Carole Cadwalladr and Emma Graham-Harrison. 2018. Revealed: 50 million Facebook profiles harvested for Cambridge Analytica in major data breach.

https://www.theguardian.com/news/2018/mar/17/cambridge-analytica-facebook-influence-us-election [accessed: April 02, 2018]

http://knowledge.wharton.upenn.edu/article/leveraging-customer-analytics-hotels-travel-agencies/

https://www.clickz.com/big-data-in-the-travel-industry-how-can-travel-companies-do-more-to-collect-and-use-customer-data/112872/

http://www.brickstream.com/

https://gomedici.com/8-companies-excelling-retail-analytics-north-america/

Chapter 2
CABology—What Is In It?

*I think the next [21st] century will be the century of complexity.
We have already discovered the basic laws that govern matter
and understand all the normal situations. We don't know how the
laws fit together, and what happens under extreme conditions.
But I expect we will find a complete unified theory sometime this
century. There is no limit to the complexity that we can build
using those basic laws.*

—Stephen W. Hawking

*"Unified Theory" Is Getting Closer, Hawking Predicts', interview
in San Jose Mercury News.*

(23 Jan 2000), 29A

*Give me six hours to chop down a tree and I will spend the first
four sharpening the axe.*

—Abraham Lincoln

Introduction

If you have not generated and identified value out of the cloud, analytics, and big data (CAB) Trio Wave, do not worry! This book helps you to use CABology solve real-world business problems and drive real competitive advantage. CABology merely is an optimal balance of practicing the set of strategies, techniques, methodologies to derive and communicate valuable and actionable performance measures, benefits, and insights. CABology is to [study] focus on the art and science of optimizing the business goals in delivering true value and benefits to the customer through the cloud, analytics, and big data. It will give business of all sizes a structured and comprehensive way of discovering the real benefits, usage, and operationalization aspects of utilizing the Trio Wave. If you are new to the discipline, do not worry, it will give you the strong foundation you need to get a strategic and competitive advantage to transform your businesses. If you are already into a business that relies on [full or part] the Trio Wave, it will help you to acquire crucial skills you do not have yet to achieve excellence radically.

© Springer Nature Singapore Pte Ltd. 2018
N. Upadhyay, *CABology: Value of Cloud, Analytics and Big Data Trio Wave*,
https://doi.org/10.1007/978-981-10-8675-5_2

CABology:
Arts and science of the nature and use of cloud, analytics, and big data as Trio Wave to provide value.

Some organizations know where the business processes are unchanging, well known and understood, and core to the profit and monetization ability and are in the path of CAB adoption ("GI Cloud (Meghraj) Adoption and Implementation Roadmap," 2013; Zardari, n.d., 2014). Mostly, the organizations are quite apprehensive about the CAB usage and its value to the business. The technology is changing leap and fold and providing opportunity space to all to capture and lead or to be left alone. It is a good idea to understand the business and technology space competency before embarking the CAB adoption journey. Keurig Green Mountain, Inc., formerly Green Mountain Coffee Roasters in Waterbury, Vermont, is a specialty coffee and coffee maker company. The company utilizes big data and analytics solutions to identify customers buying patterns and behaviors. The firm has 20 different brands and more than 200 various beverages. The analytics helped the firm to identify issues before they turn out to be problematic to the customers. AutoZone managed to adjust inventory and product prices for almost 5000 stores by involving big data and analytics. The company adopted the NuoDB cloud-enabled software solution which helped it to get the actionable insights without bringing down the system quickly. Ron Griffin, CIO, stated that:

> The NuoDB CDMS scales easily requires very little administration and has proved to be resilient to date.

> Source: https://www.nuodb.com

The NuoDB not only positioned itself to provide the required solution to the AutoZone but also proactively supported the CAB initiatives. Barry Morris, Co-Founder, and CEO of NuoDB mentioned that:

> Given the stature of AutoZone in their industry and their compelling vision for how the IT function can support high-volume retail transactions, we look forward to helping to enable their customer-focused, twent first-century vision.

> Source: https://www.nuodb.com

Google had developed an intelligent model that when deployed over the big data covering Web searches could identify the population infected by influenza (Aramaki, Maskawa, & Morita, 2011; Culotta, 2010; Ginsberg et al., 2009). Further, for the US population, it tried to calculate the number of people infected by influenza by relating the people's location by the flu-related Web searches. According to Google's findings almost 11% of US population have had influenza at the flu season's peak in the mid-January 2013, and the estimate was close to the Centres for Disease Control (CDC) data collection report (Culotta, 2010). Companies get a strategic advantage when utilizing the CAB solutions effectively. For example, a Colloquy partner affirms that strategic position is attained when your loyalty program brings intelligence about what your customer wants.

In the public sector, the CAB is getting recognized, and its elements are being utilized to transform cities into "smart cities," which make intensive use of data, analytics, and intelligent processes to drive decision making and services to the public. Bank of America analyzed its customer base of 50 million across all channels and interactions to provide customized offers. General Electric (GE) integrates myriad sensors to collect data from a massive stream-driven GE generator about its heat, pressure, vibration, etc., that can power up to 750,000 homes. The data collected via sensors later aggregated and analyzed with other data like weather, fuel costs, and power demand to determine the optimal performance of the generator.

The organizations entering into the CAB space must consider some guiding principles that reflect their efforts and investments and help them ensuring that those are aligned with the corporate priorities. Some statements to be looked into:

- We will set the benchmark by adopting the CAB framework.
- We will be the leaders in the CAB ecosystem.
- We will create coexist ecosystem with CAB partners, collaborators, and associates.
- CAB strategy will be out differentiator factor to promise business value in our product/service offerings.
- We will a have unified CAB platform across our all verticals.
- Our DNA will recognize and comprise of the CAB.

Large companies have the colossal infrastructure, resources, and finances but sometimes lack the flexibility to incorporate unified CAB business strategy and ecosystem across the organization. On the other hand, new business entrants are agile to embrace CAB business strategy from the starting of their existence even though they lack the infrastructure, resources, and finances. Each organization has different value proposition based on their current and future business goals and vision. Some businesses embark the incremental way of experimenting with the CAB and thus execute CAB directive locally and in small projects with a short timeline to get the confidence and learning outcomes. The businesses that are ready to take a leap and believe in transformation are engaged with the CAB adoption enterprise-wide with a long-term timeline having a motive to excel, lead, and monetize.

Framework

A formal CABology framework is described to guide businesses to understand, identify, and generate the true value of their business. It comprises of, Fig. 2.1:

- Art and Science: The CABology depends on the inclusion of the art and science discipline to understand, identify, and generate business value. It involves the creation and sharing of ideas, expressions, and feelings toward the CAB elements adoption and sourcing. Also, scientifically inclined structures, frameworks, and approaches to be deployed and experimented to verify and validate

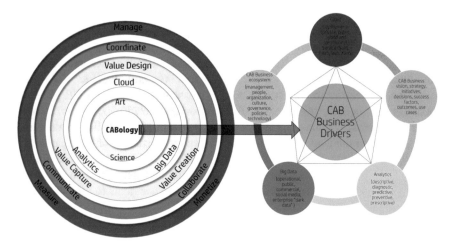

Fig. 2.1 CABology framework

the feasibility and value of the business. Leaders should possess both art and science approach and empower others to do so in embracing the CAB.

- Cloud, Analytics, and Big Data: Business to be at the forefront of utilizing the cutting-edge technology landscape and excelling at the leadership even in the complex marketing dynamics must adopt a sustainable, evolvable, and flexible framework. The cloud, analytics, and big data, a Trio Wave, already impacting the businesses and helping them to transform and attain success. The organization has to focus on the know-how of the cost, benefits, risks, opportunities associated with the CAB adoption.

  - Cloud: The cloud technology provides an intrinsic value to compute, store, and act at scale. The deployment of the cloud can be as private, public, hybrid, and community mode depending upon the need and requirements of the customers. Further, cloud consumer can utilize various cloud services which broadly classified as Software-as-a-Service (SaaS), Platform-as-a-Service (PaaS), Infrastructure-as-a-Service (IaaS), and Anything-as-a-Service (XaaS) (Larai, 2015; Vouk, 2008; Zhou et al., 2013). In particular case, an organization can adopt a hybrid level of services. The organization can be involved in cloud usage and adoption in four ways. Firstly, prepare to develop, control, manage, and deploy cloud on its own. Secondly, buy cloud services considering the deployments models (and other attributes—cost, benefits, risks, and opportunity). Thirdly, co-create the cloud services and deployment models. Finally, cooperate in utilizing the cloud services. The organization further be engaged in using cloud service as stand-alone service. For example, the organization can use single Customer Relationship Service (SaaS) service or further depending upon the requirements can use integrated services—SaaS & PaaS, PaaS & IaaS, SaaS, PaaS, and IaaS. Figure 2.2 shows the cloud computing models and strategy.

- Big data: Organization must focus on the big data technology, infrastructure, and services (BDTIS) that could benefit the organization to tap the potential value hidden inside the data that it has or could have (Greengard, 2014; Opresnik & Taisch, 2015). BDTIS to be identified to deal with the mobility (continuous, batch, and edge data), and structure (structure, semi-structured, and unstructured) dimensions of the data; to extract, transform, and load the appropriate data; to prepare and process the data, and to visualize the actionable insights. The organization can also involve in preparing, buying, co-creating, and cooperating mode to design, develop, and deploy BDTIS. Further, the associated, collaborators and partners can be generator, broker, regulator, and aggregator. Generator deals with the generation of BDTIS and data for the consumption. Broker provides appropriate third-party data and services. Regulator ensures legal, regulatory, and compliance requirement to be working with BDTIS and data. Aggregator aggregates service and data for the consumption and utility. Figure 2.3 shows big data environment.

- Analytics: The purpose of analytics is to augment, supplement, compliment, and in some situations substitute decision making and actions. The organization has to dive inside deep to find out the purpose and growth drivers to perform analytics. Six different categories of analytics—descriptive, diagnostic, predictive, prescriptive, preventive, and autonomous—can be utilized based on the need, requirement, and purpose of the organization (Fan, Lau, & Zhao, 2015; Singh & Singh 2012). Analytics' model engagement can also be done in a variety of ways—preparation of analytical model in-house; buying the analytical solutions; co-create the analytical solutions and cooperate in designing, and utilizing the analytical solutions. Further, the analytical solutions can be deployed over the cloud and run on the big data, Fig. 2.4.

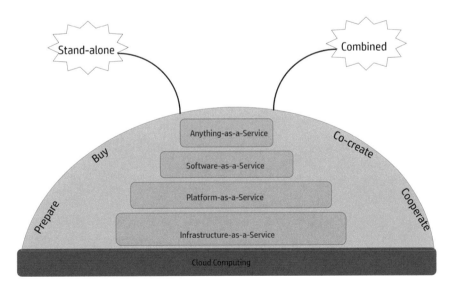

Fig. 2.2 Cloud computing business models and strategy

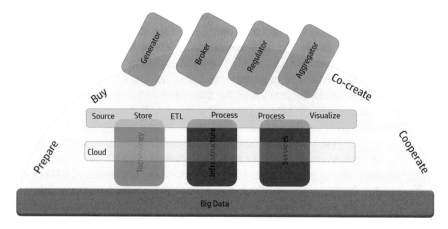

Fig. 2.3 Big data environment

- 3 V's: Organization to emerge as a leader needs to focus on three levels of values—value design, value capture, and value creation (Olsen & Welo, 2011; Opresnik and Taisch 2015; Solutions, Ceos, It, & Value, 2009; The, Of, Teradata, Data, & With, n.d.). The organization's business strategy vision, strategy, initiatives, critical success factors, decisions, and outcomes trigger the value design space. In value creation, the different drivers associated with the cloud, analytics, and big data to be considered. Finally, the value can be captured considering the complete value chain of the business—upstream and downstream. Figure 2.5 shows 3 V's model.
- 3 C's: The stakeholders need to coordinate, collaborate, and communicate to align their interests, and expectations during the CAB adoption or sourcing (Gould, 2012; Knoke, 2013). It will help them to understand any risks

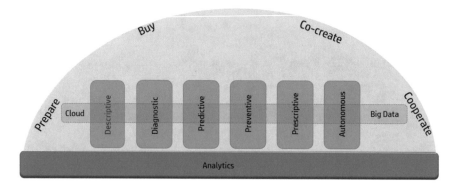

Fig. 2.4 Analytics environment

Fig. 2.5 3 V's—Value
design, value capture, and
value creation

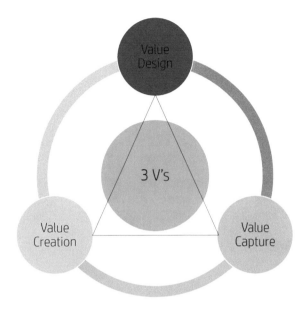

associated with the operationalization of the CAB elements—analytics, selection, and evaluation of analytical models, cloud solutions, and BDTIS to be developed, deployed, and utilized; streamline the project's timeline; prepare the right resources; identify the scope and constraints, and generate value proposition. Figure 2.6 shows 3 C's model.

- 3 M's: It deals with managing, measuring, and monetizing CAB adoption or sourcing for the business value. Managing CAB adoption or sourcing means the establishment of directives, guidelines, policies, compliance, standard operating procedures (SOPs), governance, service-level agreements (SLAs), service-level management (SLM), roles/responsibilities, frameworks, etc. ("Integrating Performance Measures in Project Portfolios," n.d.; Thaduri, Galar, & Kumar, 2015; Wu, Kumar Garg, & Buyya, 2012). To measure the CAB adoption or sourcing success, the organization needs to develop, deploy, and execute key performance indicators (KPIs), key agility indicators (KAIs), key governance indicators (KGIs), understand value and economics. Finally, to monetize the CAB adoption or sourcing organization must focus on prioritization of strategies, initiatives, business use cases, and change and transformation execution. Figure 2.7 shows 3 M's model.
- CAB drivers: The CAB business drivers are the essential elements of the CAB value proposition that help the organization to charter its business value creation pathways and action plans. At the initial phase, the organization needs to conduct feasibility assessment, planning of the engagement and determination of the primary, secondary, and other stakeholder's requirements, and service and product delivery approach. Further, evaluation of the business and IT ecosystem

Fig. 2.6 3 C's—
Communicate, collaborate,
and coordinate

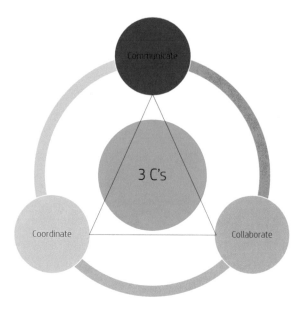

Fig. 2.7 3 M's—Monetize,
manage, and measure

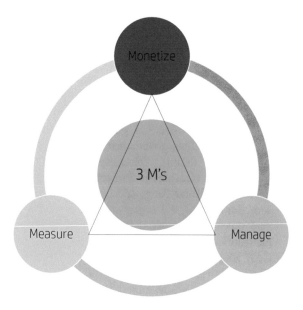

to be followed by considering aspects of management, people, process, information, and technology.

The success of CAB adoption and leadership depends upon the business leaders,
and they fall into four main categories:

Avoider: Business leader in this category avoid any challenges dealing with the inclusion, adoption, and execution of CAB-related aspects. The leader showcases behavioral, psychological, and competency blockages which in turn hamper the growth of the organization.

Resister: In this category, business leaders with much less willingness try CAB strategy and elements in their business line. Perhaps the pressure due to the environment, business ecosystem, market dynamics, and government contribute these leaders to incorporate CAB elements and aspects. They take more time to conceptualize, design, and execute CAB-related projects. They do not have confidence and motivation to embrace the CAB business strategy adoption.

Accepter: Business leaders in this category are motivated and self-driven to accept the business halogens. They understand the market and business dynamics and thus take the opportunity as and when appears. They put extra efforts to conceptualize, design, and execute CAB-related projects. Such leaders show willingness to adopt CAB business strategy and initiatives.

Seeker: In this category, the business leaders don't wait for the opportunity to embrace the technology and business landscape. Instead, they keep themselves abreast with the latest and future trends in the technology and business ecosystem. They are highly motivated and not only self-driven but motivate and lead the people and organization. The seekers create the opportunity space and are risk takers. Figure 2.8 shows CAB adoption and leadership traits.

Fig. 2.8 CAB adoption and leadership traits

Concluding Remarks

Business to be at the forefront of utilizing the cutting-edge technology landscape and excelling at the leadership even in the complex marketing dynamics must adopt a sustainable, evolvable, and flexible framework. A formal CABology framework is described to guide businesses to understand, identify, and generate the true value of their business.

References

Aramaki, E., Maskawa, S., & Morita, M. (2011). Twitter catches the Flu: Detecting influenza epidemics using Twitter. *EMNLP '11 Proceedings of the Conference on Empirical Methods in Natural Language Processing*, 1568–1576.

Culotta, A. (2010). Towards detecting influenza epidemics by analyzing Twitter messages. *1st Workshop on Social Media Analytics*, (May), 115–122. https://doi.org/10.1145/1964858.1964874.

Fan, S., Lau, R. Y. K., & Zhao, J. L. (2015). Demystifying big data analytics for business intelligence through the lens of marketing mix. *Big Data Research, 2*(1), 28–32. https://doi.org/10.1016/j.bdr.2015.02.006.

GI Cloud (Meghraj) Adoption and Implementation Roadmap. (2013), (April 2013).

Ginsberg, J., Mohebbi, M. H., Patel, R. S., Brammer, L., Smolinski, M. S., & Brilliant, L. (2009). Detecting influenza epidemics using search engine query data. *Nature, 457*(7232), 1012–1014. https://doi.org/10.1038/nature07634.

Gould, R. W. (2012). Open innovation and stakeholder engagement, *Journal of Technology Management & Innovation, 7*(3), 1–11.

Greengard, S. (2014). Big data = big challenges. *CIO Insight*, 1.

Integrating Performance Measures in Project Portfolios. (n.d.).

Knoke, B. (2013). A Short paper on innovation capability maturity within collaborations. *NGEBIS Short Papers*. Retrieved from http://www.bik.uni-bremen.de/BIK_Daten/pdf_daten/6_2013_Kno_717.pdf.

Larai, M. B. (2015). A review of service selection in cloud computing, (October).

Olsen, T. O., & Welo, T. (2011). Management & innovation maximizing product innovation through adaptive application of user-centered methods for defining customer value, *Journal of Technology Management & Innovation, 6*(4).

Opresnik, D., & Taisch, M. (2015). The value of big data in servitization. *International Journal of Production Economics, 165*, 174–184. https://doi.org/10.1016/j.ijpe.2014.12.036.

Singh, S., & Singh, N. (2012). Big Data analytics. *2012 International Conference on Communication, Information & Computing Technology (ICCICT)*, 1–4. https://doi.org/10.1109/ICCICT.2012.6398180.

Solutions, S. I. T., Ceos, W., It, T., & Value, B. (2009). Cloud computing—Business models, value creation dynamics and advantages for customers. *Computing, 15*, 1–8. Retrieved from http://scholar.google.com/scholar?hl=en&btnG=Search&q=intitle:Cloud+Computing+?+Business+Models,+Value+Creation+Dynamics+and+Advantages+for+Customers#0.

Thaduri, A., Galar, D., & Kumar, U. (2015). Railway assets: A potential domain for big data analytics. *Procedia Computer Science, 53*(1), 457–467. https://doi.org/10.1016/j.procs.2015.07.323.

The, U., Of, V., Teradata, T. H. E., Data, U., & With, A. (n.d.). Unleashing the value of the teradata unified data architecture, 1–7.

Vouk, M. A. (2008). Cloud computing—Issues, research and implementations. *Proceedings of the International Conference on Information Technology Interfaces, ITI*, 31–40. https://doi.org/10.1109/ITI.2008.4588381.

Wu, L., Kumar Garg, S., & Buyya, R. (2012). SLA-based admission control for a software-as-a-service provider in cloud computing environments. *Journal of Computer and System Sciences, 78*(5), 1280–1299. https://doi.org/10.1016/j.jcss.2011.12.014.

Zardari, S. (n.d.). Using obstacles for systematically modeling, analysing, and mitigating risks in cloud adoption, 275–296. https://doi.org/10.4018/978-1-4666-2199-2.ch014.

Zardari, S. (2014). Cloud adoption : Prioritizing obstacles and obstacles resolution tactics using AHP.

Zhou, J., Leppanen, T., Harjula, E., Ylianttila, M., Ojala, T., Yu, C., et al. (2013). CloudThings: A common architecture for integrating the Internet of Things with cloud computing. *Proceedings of the 2013 IEEE 17th International Conference on Computer Supported Cooperative Work in Design, CSCWD 2013*, 651–657. https://doi.org/10.1109/CSCWD.2013.6581037.

Chapter 3
CAB Framework—The Fabric

Every theory is a self-fulfilling prophecy that orders experience into the framework it provides.
—Ruth Hubbard (American Biologist, b. 1950)

Only a people serving an apprenticeship to nature can be trusted with machines. Only such people will so contrive and control those machines that their products are an enhancement of biological needs, and not a denial of them.
—Sir Herbert Edward Read
From The Grass Roots of Art: Lectures on the Social Aspects of Art in an Industrial Age (1955), 157. Collected in Herbert Read: Selected Writings: Poetry and Criticism (1964), 357.

Introduction

Organizations no longer need the traditional IT services in the changing world of digital business and fierce competitions (McKinsey & Company, 2011; Sánchez, Lago, Ferràs, & Ribera, 2011). They require that their IT infrastructure, whether they are the consumer or producer of the resources, should enable them to be agile, more powerful, and cost-effective. Today's business ecosystem requires that the IT infrastructure must support the change in current and future business needs and strategies. The chapter explores the CAB framework principle and concept in length. It will help organizations to build an innovative ecosystem where all resources, tools, and related applications (deep analytics or otherwise) are linked together in a unified architecture. The CAB framework (fabric) helps the organization to operationalize the value of big data. For example, Gerhard Kress, Director of Mobility Data Services at Siemens, nicely says, "We are heading towards next-generation maintenance" (Piletic, 2017). Every one like trains to operate on time as per the schedule. It put a great challenge on the train operator and to meet such requirement of time constraint and service to meet, an efficient maintenance program is critical. Siemens developed "Internet of Trains" to operationalize the value of big data in real time thus achieving a shift from reactive to predictive maintenance.

© Springer Nature Singapore Pte Ltd. 2018
N. Upadhyay, *CABology: Value of Cloud, Analytics and Big Data Trio Wave*,
https://doi.org/10.1007/978-981-10-8675-5_3

The digital oil industry by the virtue and utilization of the technology could now model virtual world of drilling, pumping, and flow (Da Silva Neves & Camanho, 2015). Consider a scenario where leakage in the pipe needed inspection or audit cycle. To perform such kind of cycle, the first-hand action involves—identification of a location, tracing of the fault, follow-up of the SOPs and specific activities, execution of trials runs, and development of the reports piled up by several pages. Nowadays, such kind of scenarios is more often transformed to "drilling to mitigating" in a digitized manner. The sensors are placed in the digital and pumping equipment's to measure, monitor, and control the drilling and pumping activities. Not only the key stakeholders can find "*what*" is happening they also able to predict the maintenance cycle required to avoid any faults and failures and minimize risks regarding the leakage and breakage of the equipment. Earlier to repair and fix faults and failures require timeline of months but with the availability of [big] data and analytic process preventive mechanisms are easily deployed resulting in saving billions of dollars to the industry.

Rolls-Royce is a pioneer in the engine manufacturing industry and produces high-performance engines for the aircraft. The Rolls-Royce engine incorporates sensors so that the information and health of the engine are available to the maintenance crew. It helps them to strategize their prescriptive and routine maintenance plan and activities. For example, the maintenance team can prepare themselves for the servicing activity before any problem occurs.

In the year 2005, the oil prices increased and a significant decision for Rocky Mountain Steel Mills ought to be taken to run the business. The company was founded in 1892 as Colorado Fuel and Iron Company, and in 1992 it has changed its name to Rocky Mountain Steel Mills. The Company specializes into steel fabrication and construction and produces a wide variety of products such as steel rail, rod, and bar, seamless tubular and customized products, wire rod, coiled reinforced bar, seamless casing and semi-furnished cast rounds. The Company though having strong leadership in the steel fabrication and construction industry forced to shut down its pipe mills in the year 2003 due to enormous price pressure. But in the year 2005 due to the variation in the oil prices, it had no choice but to rethink and decide on various strategical decisions about its leadership and pipe mills operations.

The customers, managers, supervisors, oil drillers, and critical stakeholders have shown their concern and also vocal about their interests in the pipe mills operations. Several questions need to be answered—*should the milling operation make functional? Whether to start production right away or to wait for some more time? How long to wait to gain a competitive advantage if any? Should the orders be considered now or after to be taken only after the commencement of the production?* Various factors such as prices, market, supply, demand, product constraints, benefits, and industry capacity control the decision dynamics.

The Company instead reacting to such dynamics took an informed decision way by embracing the big data and analytics pathway. The Company took a strategic decision to delay the operational activity of the mill and further delay the order process till the commencement of the production facility. The decision helped the Company to maximize its profit, market share and even increase in the stock price while avoiding the market risks.

You must be thinking that the adoption of big data and analytics journey has helped Rocky and Mountain to attain its leadership and to start its pipe mills operations. Companies like Rocky Mountain do much more to leverage the data and analytics. They have the right *vision*, their efforts are *coordinated* toward *bringing business strategy*, and they charter *business initiatives* to generate *business success factors* and *decisions*. They develop *business use cases* enabled by *management* and *technology drivers* to create *business outcomes*. Finally, they build the *right culture* and *governance* that allows them to attain *business leadership* and *business value*. The CAB strategy sets the agenda and direction of your business vision. For example, Caesar's vision was to capture value from the "*Loyalty plus service*" and realize it they concentrated their operational, technology, service and marketing activities toward customer loyalty, high-impact financial benefits, and data-driven decision makings.

Framework

A structured framework, Fig. 3.1, is proposed that prepares an organization to initiate the CAB adoption or sourcing journey.

- Vision

  The business vision is needed to

- encompass the business portfolio
- align the efforts toward addressing the best for the business
- formulate the road map, strategic and tactical action plans
- create value ecosystem

Businesses in the absence of clear business vision fail to attain the success, as the business will do everything and anything and have spillovers due to lack of the vision (Hall, Braverman, Taylor, & Todosow, n.d.). An athlete while running on the track knows the start and end and utilizes his/her strategic and tactical decision on the move to win the race. Without the knowledge of the end destination, the athlete will lose focus and have no energy to run despite having a practice run for years. So, no matter what your business has been doing for many years, if you don't set up a vision then do worry as it will let your business go somewhere but might not at

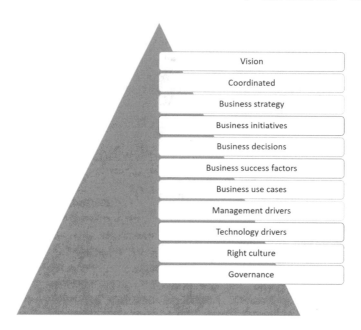

Fig. 3.1 CAB framework

the right end destination. The business vision is always a long-term and needs to be relooked based on the market dynamics, business capabilities, competitiveness, and impact. One must focus efforts to utilize the resource-intensive projects to initiate analytics efforts in generating valuable insights. One cannot focus on everything and can lead to losing sight of the business value and purpose. The enhanced customer experience was the focus of Harrah's Hotel and Casino and to achieve that they put their efforts in utilizing big data and analytics processes. Caesar's entertainment had to rethink its strategic vision to move away from opening more Casinos to using its current Casinos and customer base.

Once you agree upon a shared vision, you need to identify the strategic entry point to generate the business value. Here, you can categorically perform discussions to list down the opportunities, and a road map wherein you also allow in incremental evolution of your business vision. The entry points help the key stakeholders to focus on the activities, use cases, business initiatives, business, and technology drivers. One thing you should understand that vision should enable innovation, which may require embracing dynamism and transformation reaching the desired end state. The entry point for a cloud adoption can be vertical, horizontal, and hybrid depending upon the vision. In the vertical entry point, you may wish to put/utilize cloud service, analytics, and big data for understanding customer buying pattern during the order fulfillment process and thus getting the informed insights for increasing sales, recommending products, and minimizing inventory levels. Similarly, horizontal entry points will be helpful in covering a cross-functional aspect of the businesses. To avoid the entry point lock-in and also to get the best benefits of the entry points, one can go with the hybrid entry points.

- Coordinated

The essential ingredients of a high-performance team are the synergy due to the balanced coordination among the team members. Pick the best players of the cricket world cup team and form a new cricket team and let the team be allowed to play for a cricket match. You might be surprised to see such a game. Until unless there is a balanced synergy in a coordinated way, the team will not turn out to be a winner. The giant retailer Walmart allowed its suppliers and other vendors to use its Retail Link System to plan for products inventory, movements, and offers. The whole supply chain benefitted significantly due to the coordinated effort of the stake-holders. Procter and Gamble (P&G) moved most of the project ideas from in-house research and development projects to the collaborative projects. These projects were not conceived in-house but grew on the collaborative fronts utilizing the crowd-source strategy of designing and executing the projects. P&G also shares its data and analytics to its retail and customers through the Joint Value Creation program that helped the stakeholders to be more responsiveness and to achieve cost reduction. Connectbeam, a product used by P&G employees to share among themselves their bookmarks and searches with appropriate descriptive words that helped to deliver active social networks of coworkers to identify information swiftly.

HDFC bank in collaboration with SAS deployed an integrated CRM solution to achieve the cross-selling. It was now easy for the HDFC executives to identify whether a customer is having an active account or just the salary account. On top of that, they were also able to find whether the customer used HDFC account as a primary account. HDFC could incorporate FastTrack loan approval system, as it was possible for them to filter out the customer within few minutes who would be potential defaulters based on the analytical models, customer data, and scores. For the wholesale credit approval, segment-based profiling and scores are utilized for the loan approval.

You need to identify the coordinators, who all are going to get affected by your service, products, offerings, decisions, and actions. A business must investigate its status-quo and its upstream and downstream flow of materials, data and services to find out the primary, secondary, and other coordinators.

- Business strategy

The business strategy represents the scope of the CAB efforts and resources utilized for identifying and generating the business values. Your business strategy statement should not be ambiguous, and thus a correct short title will be helpful to understand the underline objective. The title *"Improve Supplier and Customer intimacy"* is good enough to understand what a business would like to achieve. The whole efforts to be put in connecting, understanding, expressing and realizing requirements, needs and expectations of all the stakeholders, collaborators, and beneficiaries.

In managing the organization, it is very critical to provide the needed information for every decision maker in the upstream and downstream.

- Business initiatives

Business initiatives extend and support the business strategy. Your business initiatives should not be out of sync with the business strategy. Usually, 3–5 business initiatives are recommended for each business strategy. More than that result in a bulky wish list and would not be appropriate to keep, as it won't generate any business value. The business initiatives should also be well supported by the financial and business goals against which the feasibility and success factors to be identified. A long and robust impact solution for a business problem may not be feasible to execute or implement regarding financial viability, resource utilization, or timeline. The business initiative "Increase Supplier and Customer engagement" enables an organization to look into the way to connect "upstream and downstream." The business initiatives will help the organization to understand *"what needs to be done," "what are the outcomes," "how success will be measured,"* and *"what support and service components are essential."* Examples of several business initiatives in line with the business strategy *Improve Supplier and Customer intimacy* are:

- Increase customer engagement: provide more efficient services, offer new membership plans, suggest new products and services that could delight the customer, provide geolocation-based services and offers, support more reward and loyalty points, etc.
- Increase supplier engagement: provide details on inventory, customer's buying pattern, and the likelihood of buying products and services, customer's feedback about the services and products, etc.
- Connect supplier and customer: achieve cross-sectional sharing of supplier and customer data, attain new discounts and bulk order plans, generate and share deeper insights that could benefit both customer and supplier.

- Business decisions

Business decisions are vital as they contribute significantly to the success of the business initiatives. These decisions can influence functional and cross-functional business domains and processes.

Examples of several business decisions in line with the business strategy *"Improve Supplier and Customer intimacy,"* and business initiatives are:

- Increase customer engagement: whether to provide *"plastic-free"* or *"plastic"*-based memberships, whether to offer 10x rewards or 5x rewards points to move the level of customer membership from *"silver"* to *"gold,"* etc.
- Increase supplier engagement: whether to provide *"online"* or *"offline"* details on inventory; whether to give details of every transaction (daily/weekly/

monthly) of customer's buying pattern and the likelihood of buying products and services; whether to share weekly or fortnightly consolidated customer's feedback about the services and products, etc.

– Connect supplier and customer: whether to concentrate functional or cross-sectional sharing of supplier and customer data, whether to attain new discounts and bulk order plans in an incremental fashion or fixed way

• Business success factors

Business success factors help the organization to achieve "*what needs to be done,*" "*what are the outcomes,*" "*how success will be measured,*" and "*what support and service components are essential.*"

Examples of several business success factors in line with the business strategy "*Improve Supplier and Customer intimacy,*" business initiatives, and business decisions are:

– Increase customer engagement: whether to provide "*plastic-free*"- or "*plastic*"- based memberships, whether to offer 10x rewards or 5x rewards points to move the level of customer membership from "*silver*" to "*gold,*" etc.

Increased engaging Loyalty clientele: develop an understanding of customer's behaviors, buying patterns, interests, likings, experience, and accessibility.
Increased retail membership: develop an understanding of customer's financial status, portfolios of shopping cards, accessibility, and openness to membership.
Increased online discounts: develop an understanding of customer's navigational and online cart behaviors, buying patterns, interests, likings, experience, and accessibility

– Increase supplier engagement: whether to provide "*online*" or "*offline*" details on inventory; whether to give details of every transaction (daily/weekly/monthly) of customer's buying pattern and the likelihood of buying products and services; whether to share weekly or fortnightly consolidated customer's feedback about the services and products, etc.

Increased reliable suppliers: Develop an understanding of supplier's behaviors, service patterns, experience, and accessibility.
Increased just-in-time products and services: Develop an understanding of supplier's financial status, portfolios of shipping products and service offerings, accessibility and openness to quick service and products offerings.
Increased quality products and services: Develop an understanding of supplier's product and service quality, and accessibility of reshipping of faulty service and products.

– Connect supplier and customer: whether to concentrate functional or cross-sectional sharing of supplier and customer data, whether to attain new discounts and bulk order plans in an incremental fashion or fixed way

Increased engaging Loyalty clientele: Develop an understanding of customer's and supplier's behaviors, buying patterns, service patterns, experience, and accessibility.

Increased user experience with products and services: Develop an understanding of customer's and supplier's experience of shipping/servicing and utilization of the products and services

Increased accessibility of products and services: Develop an understanding of customers and supplier's approachability, accessibility, and availability of the products and services.

- Business use cases

A business use case defines the scope and elements involved in achieving the business initiatives in context. Each business use case covers various stakeholders' perspective of interacting with the service, products, and offerings. The primary usage of the business use case is to provide an understanding of collaborators, partners, and stakeholders in driving the business. A business use case that you might consider as secondary might be considered as a primary business use case to other organization. For example, an insurance company will use an insurance software to provide a central business use case of giving valuable insurance offerings to its customers. For them, insurance software development is the secondary and an external use case which might be the primary use for an insurance software development company. Each business initiatives will have several business use cases. Each business use case needs two elements—input from an actor and a business rule. An actor is defined an object which/who is involved in stimulating the trigger to the business rule. The business rule has a business process that upon triggering may result into activating other use cases or support other actors toward the use case goal.

Examples of several business use cases in line with the business strategy "*Improve Supplier and Customer intimacy*," business initiatives, business decisions, and business success factors are:

- Order Process: "customer order details, choices, and reviews are shared with the supplier for more informed actions"
- New product development: "customer buying pattern, profile, segments, choices, and review to be shared with the supplier for the new product development. Customer will also get the chance to test the beta version of the service or product"

- Management drivers

The management is responsible for driving and leading the organization. It is also required to work in synergy to make sense out of the situations faced by the organization, make informed and actionable decisions, and develop strategic and tactical action plans toward the value and goal of the organization. Managers must act responsibly and ethically to perceive organizational challenges and identify optimal ways to allocate financial, human, technical, and skilled resources to coordinate, collaborate, and achieve success. Managers must also from time to time reflect on the outcomes and not only to outputs. They must also be responsible for integrating organization's technology and business strategy. A substantial part of management is to develop innovative and creative artifacts with new knowledge, insights, and information.

Examples of several management drivers in line with the business strategy "*Improve Supplier and Customer intimacy*," business initiatives, and business decisions are:

– Increase customer engagement, select customer engagement software, integrate cross-functional units, score each customer, generate segment-based rewards and recommendation, and develop products ratings
– Increase supplier engagement, select supplier engagement software, incorporate cross-functional vendor/suppliers units, score each supplier (based on product quality, delivery timeline, Quality of Service (QoS), business support, etc.), generate support-based rewards, develop products usage and recommendation plan
– Connect supplier and customer, integrate supplier and customer engagement software, facilitate the flow of information, insights, and analytics upstream and downstream.

• Technology drivers

Technology is an ultimate valuable tool to drive the business vision, strategy, initiatives, decisions, use cases, goals, etc. Management invests immensely in the technology to get the economic viability and value to their business. Most of the time the primary concern for the management to use technology is to ensure that it delivers the genuine business benefits. The value of the technology drivers is in large part, determined by the extent to which it contributes to management decisions, insightful actions, increased productivity, generate value, and efficient operations. Technology drivers are a valuable instrument in developing the value for the businesses.

Examples of several technology drivers in line with the business strategy "*Improve Supplier and Customer intimacy*," business initiatives, and business decisions are:

– Increase customer engagement, implement and deploy customer relationship management (CRM), use SAP, integrate CRM and SAP over the cloud for generating informed and actionable decisions and managing customers, generate dashboard at the retail frontline to provide dynamic access to the frontline manager for providing just-in-time offers/discounts.
– Increase supplier engagement, implement and deploy vendor relationship management (VRM), implement and deploy materials management (MRM), deploy SAP, integrate VRM, MRM, and SAP over the cloud for generating informed and actionable decisions and managing vendors, generate dashboard at the vendor frontline to provide dynamic accessibility to vendors for providing just-in-time offers/discounts and quality of service and products.

- Connect supplier and customer, implement supply chain management system (SCM), and integrate it with cross-functional technology tools on-premise or over the cloud facilitating the flow of information, insights, and analytics upstream and downstream.

Organization needs to understand their stand and strategy to adopt cloud to drive and support the business strategy vision. The key stakeholders are required to keep in mind following simple but profound considerations:

- Roles: It deals with the aspects and the role that stakeholders play considering its association with the cloud

  Are they the consumers of the cloud service?
  Are they the producers of the cloud service?
  Are they the broker of the cloud service?
  Are they the regulator of the cloud service?
  Are they the aggregator of the cloud service?
  Are they interested in the mixed proposition of the cloud service?
  Are they interested in the hybrid proposition of the cloud service?

- Service types: It deals with the kind of service stack that the stakeholders are interested in getting the competitive edge. Higher the level of the service in the service stack, more structure and standard the processes are aligned with the business strategy vision.

  IaaS: It represents the bottom-most layer in the cloud service where the infrastructure control and management operations are provided to the stakeholders.
  PaaS: It represents the layer that hides the complexity of the IaaS layer and provides the development and deployment platform to the stakeholders.
  SaaS: It deals with the software applications, for example, CRM and ERP, accessible through cloud over different machine configurations.
  XaaS: It represents the specialized and standardized business process, for example, employee management and procurement service, accessibility through cloud service. It is the highest layer and abstracts the complexity of the other layers.

- Deployment models: It deals with the type of service deployment to be used by the stakeholders.

  Private: It represents the cloud service being consumed and produced is at the same organization, although the third party can manage the service.
  Public: It represents that the cloud service being consumed and produced is at the different organization, although the asset ownership requires delivering the cloud service typically remains with the service provider.
  Hybrid: It deals with the combination of private and public cloud service, and no restriction on the combination of service types, consumers, and providers.

Community: It deals with the cloud service especially for a certain community, for example, health care group or journalists. The service can reside on or off premise, and the asset ownership is with the organization or managed by the third party.

Data sources are also an essential element for a CAB framework as it drives the value when controlled, managed, processed, and analyzed properly and timely. Stakeholders need to identify internal, external, and third party data sources that could be utilized for the business initiatives. Some data sources are of great importance:

- Customer data: It deals with the data related to customer's demography, qualification, experience, education, and personal, psycho-demographic, behavioral.
- Transaction data: It deals with the data related to customer's shopping transaction including purchase, return, and gifts
- Contact data: It deals with the data associated to customer's contact information which includes email id, SSN, phone number, address, and social media information.
- Marketing data: It deals with the data related to marketing campaigns, conversion, leads, segments, and failures
- Log data: It deals with the data associated to customer's event and transaction logs which includes telephone logs, email logs, browsing logs, communication logs.
- Email communication data: It deals with the data related to email communication inside and outside of the organization.
- Social media data: It deals with the data related to social media communications.
- Beneficiaries: It represents the end and business users who are utilizing your organization's service and thus are the prime beneficiaries. They are responsible for valuing your service only if that appeals to them, fulfill their requirement, and delight them.

What services are being used?
Who are the end users?
Are there any cultural or bilingual considerations exist?
Is there any category of workloads?
Are there any regulatory requirements?

- Right Culture

Organizations that succeed with CAB usage *do not do* so because they have the better accessibility of cloud, analytics, and big data solutions. Rather, organizations succeed because they have better leadership that creates the valuable and right culture for the organization to grow, set, and articulate realistic and valuable goals, and puts everyone *on board*. Each organization has a unique culture, a way of doing things, and value system. In the absence of the right culture, it is very challenging and most of the time difficult to attain the business strategy vision (Rodríguez Ortega, 2013; SAS, n.d.).

Examples of several cultural aspects in line with the business strategy *Improve Supplier and Customer intimacy*, business initiatives, and business decisions are:

- Increase customer engagement: consider the customer as a valuable asset, respect customers feedback and critics, facilitate customer to open up to share their interests, liking, wish list, and service requirements.
- Increase supplier engagement: consider supplier as a valuable asset, respect supplier feedback and critics, facilitate supplier to open up to share their experience, viability, and product portfolio and service (SLAs) requirements.
- Connect supplier and customer: securing supplier and customer's interest in sharing the upstream and downstream flow of information.

Hofstede explained six elements of the culture dimensions (Hofstede, 1984) that impact the organization performance which can be very well adapted for the CAB framework. The culture dimensions are helpful for the organizations to understand the capability and capacity to embrace the CAB elements to execute business strategy vision.

Individualism index (IDI): The IDI deals with the way that the CAB adoption is appreciated and embraced by an individual or by collective group. The higher the index, the lower is the motivation among the group to work with CAB elements.

Power distance index (PDI): The PDI defines the level of involvement of the employees of the organization to accept the CAB into their business vision. The higher the index, the lower is the involvement of few groups, community, or leaders are interested in adopting CAB framework. It might be due to the lack of enterprise-wide interest and a common goal.

Uncertainty avoidance index (UAI): The UAI deals with the acceptance of the CAB not only in functional but also cross-functional domains and services. The higher the index, the lower is the involvement of few groups, community, or leaders are interested in adopting CAB framework. It might be due to the lack of awareness, benefits, effectiveness, and utility.

Masculinity index (MAI): The MAI deals with the way that employee in the organization considers the promotion, motivation, usage, and appreciation of CAB elements. The higher the index, the lower is the support of employees to other employees in their CAB enterprise-wide projects. It may happen due to the lack of appreciation, rewards, appreciation assertiveness, and achievement.

Long-term orientation index (LOI): The LOI deals with the way that organization's functional and cross-functional projects embrace the CAB enterprise-wide. A short-term benefit is not feasible and do not provide long-lasting value to the organization. The lower the index, the higher is the adaptation and circumstantial projections. Initially, a short-term goal that might execute the proof of concept of validating CAB benefits can be adopted, but for a more significant impact and ROI, the long-term orientation is required.

Indulgence index (INI): The INI deals with the way that organization's functional and cross-functional projects being controlled, managed, and executed by high-performance teams. Every employee of the organization believes in contributing toward the CAB goals and participates in the success of the business vision. The higher the index, the higher is the utility and execution of CAB projects. The team can control the dynamics of the changing technology.

- Governance

Governance is critical to the success of your CAB framework adoption. You need to verify and validate the feasibility and value of the CAB framework in your business environment and also the change required to do so. Governance makes people aware of their roles, responsibilities, policies, strategies, actions, authorities, and accountabilities (Morabito, 2015; Provan & Kenis, 2007). Ross, Weill, and Robertson (2006) have described IT governance in their book Enterprise Architecture as Strategy—"specifying the decisions rights and accountability framework to encourage desirable behavior in the use of IT." Nonetheless, you can apply the same principle to the CAB governance enterprise-wide as "specifying the decisions rights and accountability framework to encourage desirable behavior in the use of CAB [framework]." The interesting point is that the management drives the execution of the governance, but management is not governance. Management deals with the action and implementation of the decision while governance covers aspects related to decision rights, initiating, conceiving and making decisions, and establishing accountability.

Concluding Remarks

Today's business requires that the IT infrastructure and business ecosystem must support the change in current and future business needs and strategies. To achieve such a challenging task, CAB framework is proposed. It will help organizations to build an innovative ecosystem where all resources, tools, and related applications (deep analytics or otherwise) are linked together in a unified architecture. The CAB framework (fabric) helps the organization to operationalize the value of the Triology.

References

Da Silva Neves, A. J., & Camanho, R. (2015). The use of AHP for IT project priorization—a case study for oil & gas Company. *Procedia Computer Science*, *55*(Itqm), 1097–1105. https://doi.org/10.1016/j.procs.2015.07.076.

Hall, R. E., Braverman, J., Taylor, J., & Todosow, H. (n.d.). The vision of a smart city.

Hofstede, G. (1984). Cultural dimensions in management and planning. *Asia Pacific Journal of Management*, *1*(2), 81–99.

McKinsey & Company. (2011). Big data: The next frontier for innovation, competition, and productivity. *McKinsey Global Institute*, (June), 156. https://doi.org/10.1080/01443610903114527.

Morabito, V. (2015). Big data and analytics: Strategic and organizational impacts. *Big Data and Analytics: Strategic and Organizational Impacts*, 1–183. https://doi.org/10.1007/978-3-319-10665-6.

Piletic, P. (2017). Are major optimization opportunities hiding in your business data? Retrieved from https://www.smartdatacollective.com/major-optimization-opportunities-hiding-business-data/.

Provan, K. G., & Kenis, P. (2007). *Modes of Network Governance: Structure, Management, and Effectiveness*, 229–252. https://doi.org/10.1093/jopart/mum015.

Rodríguez Ortega, N. (2013). Digital humanities, digital art history and artistic culture: Relationships and disconnections. *Artnodes, 13*(13), 16–25. https://doi.org/10.7238/a.v0i13.2017.

Ross, J. W., Weill, P., & Robertson, D. (2006). *Enterprise architecture as strategy: Creating a foundation for business execution*. Harvard Business Review Press.

Sánchez, A., Lago, A., Ferràs, X., & Ribera, J. (2011). *Management & Innovation Innovation Management Practices, Strategic Adaptation, and Business Results: Evidence from the Electronics Industry, 6*(2).

SAS. (n.d.). Building a strategic analytic culture. *White Pap*.

Chapter 4
CABonomics—The True Value

The time to repair the roof is when the sun is shining.
—John F. Kennedy,
[State of the Union Address January 11 1962]

There is one and only one social responsibility of business–to use it resources and engage in activities designed to increase its profits so long as it stays within the rules of the game, which is to say, engages in open and free competition without deception or fraud.

—Milton Friedman

Introduction

Businesses tend to be apprehensive to adopt CAB framework in their organization. It is not because they don't have the financial support, but they are not comfortable and aware of identifying and getting the business value. Even though the CAB adoption is on rising and businesses are attempting to use the cloud, analytics, and big data to drive their business values, you should understand the financial and economics of the CAB adoption. The chapter will cover the financial and economics consideration that will help you to take initiatives before you communicate and incorporate CAB in your current and downstream projects. The business value analysis must consider the time to market (TTM), reduction of capital and operational expenditures (CAPEX), optimum workload and resources move, adds, changes (MAC), reduction of time to implement (TTI), the total cost of ownership (TCO) of the CAB framework (Azis, 1990; Martens, Walterbusch, & Teuteberg, 2012; Saaty, 2010; Vouk, 2008; Wu, Buyya, & Ramamohanarao, 2016).

In his classic book "Competitive Advantage: Creating and Sustaining Super Performance," Porter (1998) explained the utility and benefits of the value chain. Porter's work stimulated the usage, utility, and effectiveness of optimizing the support and operational functions covering aspects of upstream

© Springer Nature Singapore Pte Ltd. 2018
N. Upadhyay, *CABology: Value of Cloud, Analytics and Big Data Trio Wave*,
https://doi.org/10.1007/978-981-10-8675-5_4

and downstream to bring products to the market. In short, the firm's functional portfolio has two building blocks comprising of primary and secondary activities. The primary activities include inbound and outbound logistics, operation, service and sales and marketing. The secondary activities are the support activities that make the delivery of the primary activities possible and include organization infrastructure (administration and management), human resources (employee recruitment, management, hiring, engagement, and training), technology (involving improvisation of the products and the production process), and procurement. The primary and secondary activities together help the organizing to gain and generate direct and indirect values thereby contribute to producing the value creation.

In the organization, each of the value creation areas is touched or will be touched upon by the adoption of the CAB framework. Porter explained that there is an increase in the economies of scale with the operating efficiency and utilization support. A proper analysis of the CAB value chain and application of accounting, economics, and financial principles will show that adoption of CAB can influence operating efficiency and utilization support and thus achieve the economies of scale.

The supply and demand of product and services are the most fundamental economic principles (Ashtiani & Bosak, 2013; Seuring, 2012). Supply refers to the quantity of the items needed to be supplied and demand means the price one is willing to pay to obtain the item. The organization needs to understand the business cases for the adoption of CAB framework. For example, the same analytic process can be applied to different data sets for various clients. Also, the same information can be sold (licensed) and delivered to varied users. By this virtue of usage of big data and analytic processes, it is very challenging to get the equilibrium cost of such services. The supply, demand, and pricing of the CAB elements are in part determined by its utility. In a very trivial way, economists define utility as a measure of pleasure or happiness. A marginal utility refers to the way that one can get an additional satisfaction by consuming additional unit of good or service. The law of diminishing marginal utility refers to the fact that each subsequent consumption of a service or a product results in less utility than the prior consumption. But, usually interacting with the CAB framework, additional consumption of [big] data and analytic process might increase the utility, and thus the law is not "as applicable" for the CAB framework. Also, one may find having a high increase in the marginal utility of the "original content" and the [big] data utilized for the business use cases. In many cases, publishing the same information to the analytic processes does not count into an increase in marginal utility. One must notice that in the IOT environment, solution architects need to consider the marginal utility law in transmitting and sending the [streaming] data and thus most of the time edge analytics is preferable to avoid the overflow or diminishing value data.

Architecting for Optimizing CAB Utility

One can use three specific considerations when adopting CAB framework to architect the enterprise-wide system to avoid diminishing value of [big] data and processes, see Fig. 4.1.

- Transmitting distinct data: In such type of transmission, distinct data need to transmit to avoid the repeat data management and processing cost. It is better to transmit the data only when the state of the data is changed. For example, in IoT devices the data is sent when there is an update in the state of the device.
- Transmitting differential data: In such type of transmission, differential data need to transmit to avoid the repeat data management and processing cost. For example, it is always useful to transmit differential backups, data leakages.
- Transmitting derivative data: In such type of transmission, derivative data need to transmit to target particular use case and a segment of users. For example, it can be used to publish a revision of a book or an article or to contact specific segment of the customer with a potential and target recommendation.

Organization to account for the benefits and value of the CAB adoption needs to understand the concepts of economics, finance, and metrics. One can debate on measuring the value of overall business performance by balance sheets governing profits and losses. Though it seems to be easy to gauge the overall business

Fig. 4.1 Architecting for optimizing CAB utility

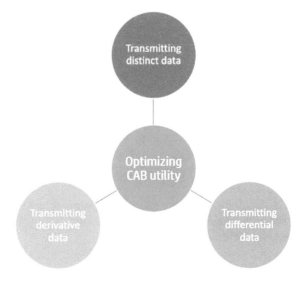

performance, it is not a silver bullet whereby using it you can get the value of your CAB adoption. The [big] data and analytics do not have space on the balance sheet, and thus leadership pays less attention in deriving the actual business value. There exist direct and indirect metrics associated with the CAB adoption. Some are general and applicable to business and IT performance.

The indirect method of monetizing CAB adoption can include operational efficiencies, reduce costs and risks, improve productivity, develop new products, services, and markets, and build new networks, and strengthen existing networks. The organization must pay attention to the fact that the monetization from the CAB adoption will be materialized only if the organization can measure its economic benefits. As most of the regulatory bodies—generally accepted accounting principles (GAAP), American Institute of Certified Public Accountants (AICPA), and Financial Accounting Standards Board (FASB)—do not consider [big] data and analytics as the capital asset, the business must take an advantage of CAB framework elements to incorporate value into its business models. The CAB has an intrinsic economic value, and business must prepare itself monetize this value.

The economic attribution will help the organization to measure the economic gains and also to record the measurement of the net value of the CAB assets. The *CABonomics* deals with the art and science of human choice and behaviors concerning the production, distribution, consumption of scarce data, cloud, and analytic resources. The key to generating measurable economic value with CAB assets involve an understanding of direct and indirect metrics. Mostly, organizations work reactively and thus loose opportunities to lead and sustain the global competitiveness.

Facebook's filing of S-1 with US Securities and Exchange Commission (USEC) has indicated reported assets of worth $6.6 billion and predicted a post-IPO market cap worth of $75 billion. Facebook had close to $68 billion non-reported information assets, that means at that time each user worth cost of $81 (Laney, 2012), and now this figure is quite appealing which is more than $200 per user (Laney, 2018).

Asset

Business leaders need to figure out the place for a CAB as an asset, more specifically to make an informed and actionable decision toward contributing to the business vision and value. In the typical nomenclature, asset is merely meaning simply "something of value." But there is more associated with it. Let us consider some of the definitions of the asset, and then we will explore its relationship with the CABonomics.

Cambridge dictionary (2014): "a useful or valuable quality, skill, or person"; "something valuable belonging to a person or organization that can be used for the payment of debts."

Wikipedia (2018): "In financial accounting, an asset is an economic resource. Anything tangible or intangible that can be owned or controlled to produce value and that is held by a company to produce positive economic value is an asset."

Merriam-Webster (2003): "the property of a deceased person subject by law to the payment of his or her debts and legacies," "the entire property of a person, association, corporation, or estate applicable or subject to the payment of debts," "an item of value owned."

Collins (Gavioli, 2005): "Something or someone that is an asset is considered useful or helps a person or organization to be successful."; "The assets of a company or a person are all the things that they own."

FASB (1985): "Probable future economic benefits obtained or controlled by a particular entity as the result of past transactions or events."

IASB (2016): "is a resource controlled by the entity as a result of past events and from which future economic benefits are expected to flow to the entity."

It is to be noted that CAB elements covering [big] data and analytic processes meets the criteria of an intangible asset as defined by accounting standards. Some of the critical attributes associated with the intangible assets defined by the International Accounting Standards (IAS, 2014) are:

- Lacking physical substance
- Projecting future economic value (such as revenue, reduce risks and costs)
- Control and manage (ownership and right to use and control the asset)
- Identifiability (capable of being separated and sold, transferred, licensed, rented, or exchanged, either individually or together with a related contract)

Indirect Metrics

A robust total cost of ownership metric is valuable when an organization is investing in the IT resource whether to use it for customer-centric portfolio or to utilize it for internal operational systems or business support systems.

The total cost of ownership (TCO): TCO analysis can be a time-consuming bit in case done in a timely and an effective manner brings competitive advantage, optimum efficiency, and reduced costs (Martens et al., 2012). The TCO in simple terms represents the total of all associated costs covering aspects of purchase, ownership, usage, operational maintenance of a particular entity. As with CAB adoption, there are several costs such as

- ownership
- maintenance, service, operational, ongoing support, bug fixes, upgrades and case escalations
- Power to run and cool servers, disks, processors, and network infrastructure
- support activities—to move, add, and change the configuration, resource and infrastructure and workloads

The investment types can be treated as expensed or capitalized. Items such as disks, servers, network routers, switches, computer room air chillers (CRAC) are considered as fixed assets and are capitalized and have depreciation life cycle (over useful life cycle). The depreciated asset also needs to be included in the TCO analysis.

TCO analysis:
  Let's take a disk storage unit that costs around $2000,000 and has a useful life of 4 years. A straight-line depreciation method, results in charge of $500,000 per year. Also, considering maintenance cost ($500,000 annually), disk physical footprint cost ($10,000 annually; cost related to space usage, power and cooling costs), MAC cost to provision storage facility to clients need a full-time engineer ($150,000). We can conclude that the annual TCO is $53660,000 and over the life cycle is $21440,000.

$$\text{Straight-line method} = \text{Depreciable amount}/\text{Useful life}$$

Item	Annual cost	Four-year cost
Disk storage	$5000,000	$20000,000
Maintenance	$200,000	$800,000
Facility foot print	$10,000	$40,000
FTE labor	$150,000	$600,000
Total	$53660,000	$21440,000

A TCO calculation is relatively easy for example above, but to measure TCO for a complete data center, server farm, business line considering CAB elements can be a decisively more complicated exercise. Also, bullwhip effect, virtualization, and multitenancy play a great role in contributing to the dynamics of the TCO.

The organization needs to maintain costs associated with functional, cross-functional analytical processes, [big] data usage services, and cloud service assets throughout the organization. Some offsets value regarding consideration of workloads, migration of workloads, load balancing to achieve performance is of utmost important by which an organization can get an estimate of the TCO. Some attributes of the CAB elements that contribute to the TCO are:

Availability: It represents one of the critical requirements for the CAB performance. The CAB elements need to be available to provide the desired service. Whether the service is for using analytical processes on big data hosted on the cloud or the service to store and manage big data on the cloud must be available regardless of whether its users are internal or external. The availability of service refers to the amount of time a service is available to its users for its usage in a given time window. For example, service to store big data on a cloud is available for 24 × 7 in a week which means a user can access the service to store or manage data anytime in a day/week.

The downtime of the service depends on the percentage of 9s involved in the service availability. It may be seen that the service that is available with "0.99999" performance has a downtime of 5.256 min annually while a service that is available with "0.9999" performance has a downtime of 52.56 min annually, Table 4.1. It has a serious implication on the cost that is to be incurred by the company during the uninformed/unplanned outage or downtime.

For example, if your organization has an application that runs on the cloud and provides travel booking service and processes $2 million of booking orders in an hour, then unavailability of the service for 5.26 min will cost the business $175200, Table 4.2.

Not only has the revenue-driven application/service impacted the business. Non-revenue drive application/service also affects the brand value of the company. An informational dashboard or a Website if down will not serve the purpose and can cause uneasiness across its user base and contribute toward a loss of trust and brand. A different business cost is associated with the type of environment and users associated with the environment. For example, a development environment does not need to be present 24 × 7 a week, a whole month, and a complete year but somewhat a working week time between 9:00 a.m. and 5:00 p.m. is required for the availability of the service. A precise and targeted service-level agreement is necessary to be drafted and followed across the service execution.

Time to Market (TTM): It is another important metric that makes an organization competitive. An organization that develops new products and services improve current product and service portfolio on the timely basis using CAB framework ought to be on the path to achieving success. The CAB framework and associated elements are prone to get affected by the people participation and bull whip effect thereby increasing TTM and costs. The organization having higher TTM face a challenging task to compete in the market as it also add-into the loss of the customer base, loss of trust and loss of stock market price. A four-quadrant view is presented in the Fig. 4.2. The 1st quadrant deals with the low time to market having a product with poor quality. No organization wishes to be this quadrant. Most of the companies fall into 4th quadrant where the time to market is high, and the quality of the product is also high. Only a few companies have their presence in the 3rd quadrant where time to market the service and product is low, but the quality of the service and product is high. Organization to be present in the 2nd quadrant is a disaster for a company to take much time to produce a poor quality of service and products.

Opportunity Costs: It is the simple cost associated with the choices when another choice is chosen. For example, a company wishes to execute two projects related to the CAB—"project A" and "project B." The net returns of "project A" is $0, and net returns of "project B" is $17500. When company chooses to execute "project A," then the opportunity cost for the company is $17500.

Table 4.1 Service availability in calendar hours

Service hours per the calendar year	Service availability	Minutes of service downtime
525,600	0.99999	5.256
525,600	0.9999	52.56
525,600	0.999	525.6
525,600	0.99	5256
525,600	0.95	26280
525,600	0.9	52560

Table 4.2 Cost to business for the service unavailability in calendar hours

Travel booking order		Travel booking order service downtime (in min)	The cost to the business
Service hours per the calendar year	Service availability		
525,600	0.99999	5.256	$175200
525,600	0.9999	52.56	$1752000
525,600	0.999	525.6	$17520000
525,600	0.99	5256	$175200000
525,600	0.95	26280	$ 875999999
525,600	0.9	52560	$1752000000

Fig. 4.2 Time to market versus quality

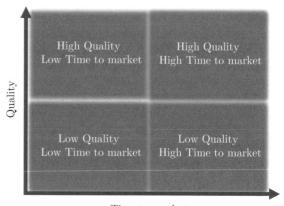

Time to market

Service-level agreements (SLA): It is a critical tool in the hands of consumer and producer of a service. The SLA sets the expectations and liability if the expectations are not met, Fig. 4.3. The CAB framework must incorporate the SLAs. For example, the SLA for the cloud storage service to store the big data must be outlined precisely regarding service availability (five 9s or four 9s). Google provides refund policy for its service agreement. Similarly, Amazon's EC2 SLA outlines target a 99.5 service availability during a service year.

Fig. 4.3 Service-level agreement

Direct Metrics

The list of direct metrics for the finance is too long, but in this section, we will discuss the select few metrics necessary for an organization to consider to adopt CAB framework. The metrics used across the enterprise-wide CAB adoption will have a common effect in understanding the values, benefits, and effectiveness. The chief finance officer (CFO), program management officer (PMO), human resource management officer (HRMO) must adhere to some common set of performance metrics.

Payback Method

It is the simplest method used by the companies to understand the value of the investment in the business. One should know in advance, by when and how much time is needed to get the money back after the service and product is functional or open for use by the customers. A shorter payback period deemed to be considered good by the organization as compared to the product or service that take a longer time for the payback.

> A company "Xees" decides to put its procurement process in the cloud. The cloud service technology promises to increase the procurement process to process the order thrice as quickly as the old on-premise system. Also, the cost to the company to process such order will be less as the cost to maintain and run those systems is not under the purview of the company. Leaving aside such costs, consider that on an average with the old on-premise system the "Xees" processes $300,000 orders in three months. The new cloud service technology, which costs around $10,000, processes the same amount of orders in 1 month. The payback for the company is fast within first two weeks.

Net Present Value

It is the method by which the organization can consider the time value of money (TVM) and the discount rate for its investment. TVM deals with the concept to have more value for the money in the future. The discounted rate measures the present value of the investment. Typically, to estimate NPV, the company utilizes the discounted rate which is predetermined by the company as it's – hurdle rate or weighted average cost of capital (WAAC). Due to the discounted rate, the future value of the investment is assumed to be more significant at the end of a period. Conversely, the present value of the investment is less than its value in some predetermined future period.

NPV

The future value of a dollar in one year is $1.07 considering the interest rate of 7 percent.

$$\text{Future value} = [\$1 + (.07 * \$1])$$
$$= \$1.07$$

Conversely, the present value of $1 from one year now with a discount rate of 7 percent is:

$$\text{Present value} = \$1/\$1.07$$
$$= 0.934$$

The NPV describes the justification for the investments and capital expenditures. In situations like adopting the CAB framework, more specifically contemplating the build or buy decisions such as building the private cloud or buying the public cloud service/resources, NPV analysis allows decision makers to choose an adoption that meets company finance and economical requirements.

Return on Investment (ROI)

It is the most common tool to justify the investment in the CAB framework and to build an appropriate ecosystem. It is a financial ratio that helps the organization to understand its profitability. The ROI is very easy to calculate, and the formula is:

$$\text{ROI} = (\text{Gain from investment} - \text{Cost of investment})/(\text{Cost of investment})$$

ROI can be used in conjunction with NPV where the present value of all gains can be considered alongside all costs of investment. The formula for the same is:

ROI of Cloud Service

A company "Xees" expects to invest $30,000 in a cloud travel booking service and support for one year, and project earnings of $150,000 after the first year as a result of it's investment, then it is very easy to identify the projected ROI as:

$$\text{ROI} = ((150000 - 30000)/30000) * 100)$$
$$= 400$$

> The projected ROI is quite high. Nonetheless, you should also compute ROI without the adoption of the cloud service and assess the projected benefit after one year and in an incremental fashion for next three years to get a clear picture about its profitability measure.

Concluding Remarks

The supply and demand of product and services are the most fundamental economic principles. The organization needs to understand the business cases for the adoption of CAB framework. Three specific considerations—transmitting distinct, differential, and derivative date are critical when adopting CAB framework to architect the enterprise-wide system to avoid diminishing value of [big] data and processes.

References

Ashtiani, P. G., & Bosak, E. (2013). A conceptual model for factors affecting the relationship between supply chain integration and customer delivery performance. *International Journal of Academic Research in Business and Social Sciences, 3*(9), 495–505. https://doi.org/10.6007/IJARBSS/v3-i9/236 .

Asset. (2003). Merriam-Webster's Collegiate Dictionary. Merriam Webster.

Asset. (2014). Cambridge Advanced Learners Dictionary. Cambridge University Press.

Asset. (2014). Technical summary IAS 38 intangible assets. https://www.ifrs.org/issued-standards/list-of-standards/ias-38-intangible-assets/.

Asset. (2016). Conceptual Framework. https://www.iasplus.com/en/meeting-notes/iasb/2016/july/conceptual-framework.

Azis, I. J. (1990). Analytic hierarchy Process in the benefit-cost framework: A post-evaluation of the Trans-Sumatra highway project. *European Journal of Operational Research.* https://doi.org/10.1016/0377-2217(90)90059-K.

Financial Accounting Standard Board. (1985). Elements of Financial Statements a replacement of FASB Concepts Statement No. 3 (incorporating an amendment of FASB Concepts Statement No. 2). *Statement of Financial Accounting Concepts No. 6.* www.fasb.org/resources/ccurl/792/293/CON6.pdf.

Gavioli, L. (2005). *Exploring corpora for ESP learning studies in corpus linguistics.* John Benjamins Publishing.

Laney, D. (2012). To Facebook you're worth $80.95. https://blogs.wsj.com/cio/2012/05/03/to-facebook-youre-worth-80-95/.

Laney, D. (2018). *Infonomics: How to monetize, manage, and measure information as an asset for competitive advantage.* Routledge.

Martens, B., Walterbusch, M., & Teuteberg, F. (2012). Costing of cloud computing services: A total cost of ownership approach. *2012 45th Hawaii International Conference on System Sciences* (pp. 1563–1572). https://doi.org/10.1109/HICSS.2012.186.

Porter, M. (1998). *Competitive advantage: Creating and sustaining superior performance.* Free Press.

Saaty, T. L. (2010). Economic forecasting with tangible and intangible criteria : The analytic hierarchy process of measurement Ekonomsko Predvi Đ Anje Primenom Merljivih I Nemerljivih Kriterijuma : Analiti Č Ki Hijerarhijski, (1), 5–46.

Seuring, S. (November 2012). A review of modeling approaches for sustainable supply chain management. *Decision Support Systems*, 1–38. https://doi.org/10.1016/j.dss.2012.05.053.

Vouk, M. A. (2008). Cloud computing—Issues, research and implementations. *Proceedings of the International Conference on Information Technology Interfaces, ITI* (pp. 31–40). https://doi.org/10.1109/ITI.2008.4588381.

Wu, C., Buyya, R., & Ramamohanarao, K. (2016). Big data analytics = machine learning + cloud computing. In R. Buyya, R. Calheiros, & A. Dastjerdi (Eds.), *Big data: Principles and paradigms*. Morgan Kaufmann, Burlington, Massachusetts: USA. https://doi.org/10.1016/B978-0-12-805394-2.00001-5 .

Chapter 5
CABevaluation—What Is Right for Me?

*In chess, knowledge is a very transient thing. It changes so fast
that even a single mouse-slip sometimes changes the evaluation.*
—Viswanathan Anand

*When we can't determine what is art—when you get to that
point where we're not sure, that's the greatest likelihood that
we're actually experiencing something great. But I think that's
what the art world is most afraid of, because you lose that
security. Then we don't know how to assign evaluation, whether
it's cultural or otherwise.*
—Richard Phillips

Introduction

The CAB adoption or sourcing, though provide several benefits and opportunity to
the organization, is susceptible to risks, failures, and setbacks when the adoption is
materialized without considering the formal evaluation goal and exercise. In the
rapid development of technology and fierce competition, businesses have to be
agile not just to lead but even for its survival. Blockbuster, an entertainment giant,
was swiped out of business when Netflix foray into the CAB journey (Teradata
Corporation, 2014). There is no one standard way to evaluate the CAB elements as
each organization is unique and in some sense, few evaluation parameters can be
mimicked (Beloglazov & Buyya, 2010). To assess CAB elements, you need to
understand various parameters considering aspects related to—cost, benefits,
opportunity, risks, etc. A formal framework is desirable for an organization to
undertake such a valuable exercise. You might feel little anxiety by looking at the
viable choices of CAB solution to be adopted. As part of your leadership portfolio,
you would like to get assurance about your options regarding profitability, productivity, operational efficiencies, collaborations, new market penetration, and
improved innovation. Further, the assurance is extended to the inclusion of approval
of stakeholders, solution providers' creditability, and their capability to include

© Springer Nature Singapore Pte Ltd. 2018
N. Upadhyay, *CABology: Value of Cloud, Analytics and Big Data Trio Wave*,
https://doi.org/10.1007/978-981-10-8675-5_5

solutions in your business model, processes and infrastructure, and implementation of strategical and tactical decisions that required CAB adoption or sourcing.

The key stakeholders must address the following questions before undertaking the CAB solution evaluation exercise:

- How do we select cloud providers that are aligned with our business strategy?
- How do we develop, test, and deploy analytical models promising to generate business value?
- How do we control, manage, and source big data and big data infrastructure services?
- How do we ensure stakeholders' concern is addressed?

Cloud Solution Evaluation

Cloud service models could be identified that are self-service changeable. The leader should understand that the cloud adoption is not the just technology-oriented, focused goal but also vital to business compliance. As dynamics of organization keep changing due to the inclusion of internal and external forces, the cloud solution provider and solutions must be aligned with the organization demands and requirements. Amazon has emerged as the first vendor to offer a wide variety of public cloud services. Amazon cloud services include Elastic Computing Cloud (EC2), Simple Storage Service (S3), and Simple Queueing Service (SQS). Many research organizations have started utilizing Amazon cloud services as a test bed for their research experiment and case evaluations. Cloud services can achieve a different level of performance depending upon the workload generated by different system processes and applications (Gubbi, Buyya, & Marusic, n.d.; Zardari, 2014). Google Application Engine (GAE) and Microsoft Azure (MA) also provide competitive cloud services.

When considering cloud solutions, review the below checklist, Fig. 5.1, and evaluate it against the cost, opportunity, benefits, and risks parameters. Ensure that the cloud solutions are in align with the CAB business strategy and you had prioritized business use cases and identified the value proposition.

- Compliance
- Credibility
- Customer-centricity
- Expertise
- Resilience
- TCO
- Resource consumption
- Accounting
- Security
- System Support

- Service response time, throughput, and availability
- System utilization, scalability, and elasticity
- Reliability
- Compliance: It is must for an organization to adhere to the legal, regulatory, law enforcement directives, and corporate mandates. Payment Data Security Standard (PDSS) law is very stringent and designed to reduce credit card frauds. One of the measures taken by PDSS is to ensure encryption of credit card transmission across public networks (Hasan Shamsuddin, & Lopes, 2014; Paul, Pan, & Jain, 2011). Consider a situation when your organization has a business case of processing the digital payment, and you have utilized the cloud service for the processing. It is found during the audits that few cases have inconsistent encryption, what you will do? Will you blame the provider? Perhaps it is the onus of the organization to perform provider's compliance check before utilizing the services. Outsourcing and sourcing decisions are critical, and providers must know your policy and conditions before any further deliberations on pre- and post-deployment of services, training, and activities happen. The geographic location of the data may be unknown to the consumer and raise issues related to jurisdiction and privacy compliance.
- Credibility: You will not tie up the knot and sign the contract with someone who is not credible in the market (Kantarci & Mouftah, 2014; Wang, Zhang, Wang, & Qiu, 2010). The credibility of a provider is based on the services and capabilities that matter the most to the customers. It is better to verify the provider's client list and client testimonials. In some case, it is advisable to have visits to the client location and discuss with the key stakeholders about the provider and its services. It will help you to validate the client reputation and credibility. It is very challenging to work with a new entrant due to the lack of provider's visibility and reputation in the market. In such cases, you may consider some pilot is running of the services depending upon the proper business use cases that do not hamper your operationalization activities.
- Customer-centricity: Providers are considered to be valuable when they are customer focused and customer-centric (Ding, Wang, Wu, & Olson, 2017; Griffin & Hauser 1993; Straub, Kohler, Hottum, Arrass, & Welter, 2013). They can be pro-active, responsive, and adaptive. Providers will not just provide the required cloud solutions and services but also help you gain visibility, competitive edge, and market presence. Perhaps they are also flexible during contract negotiations and keep the customer first as part of their corporate vision and actions.
- Expertise: Providers with deep expertise in providing cloud solutions and services have a vital role in streamlining the consumer's business portfolio (Hahn & Knott, 2008; Jula, Sundararajan, & Othman, 2014). You should consider provider's certification, awards, recognition, and testimonials to gain confidence in the expertise that they provide. For example, a cloud provider "A" has a global accreditation of delivering SaaS service like Customer Relationship Management (CRM) in banking industry is treated superior than the one who does not have any certification.

- Resilience: Avoid relationship with the provider who is not adaptive to the market dynamics and changing needs of the technology. Also, the provider must not obstruct the business operations during merger and acquisition of services and business and must provide seamless integrations and support.
- TCO: Overshooting and exemplary costs of cloud solutions and service should be avoided. The TCO of the cloud solution and services should be performed and compared with the available choices in the market (Martens et al. 2012). Do not miss to align everything with your business strategy. In most cases, TCO analysis results into an early rejection of the cloud providers due to uncompromised cost constraint. The consideration of cost while utilization of service within and across the clouds varies should be taken into account. The cost of services between and across the cloud is more due to the involvement of the bandwidth.
- Resource consumption: When it comes to resource consumption, you must ensure that your provider provides you the service that supports optimum consumption of resources (Buyya, Garg, & Calheiros, 2011). The charges will vary depending upon the disclosure of the resource consumption.
- Accounting: The hardware and software resources are not procured and acquired upfront thus the onus of accountability and ownership is shared across the cloud solutions and service value chain (Rubinstein, 2013). You may wish to utilize SaaS CRM service from cloud provider "A" which was developed on a PaaS service provided by cloud provide "B," which inherently hosted on cloud service provider "C." You need to consider the accountability, ownership, and responsibilities of cloud service in a sophisticated cloud chain. The best way is to include the terms, conditions, legal bindings, and quality and non-quality parameters (vital to organization considerations) in cloud SLAs.
- Security: The virtualization and multitenancy introduce new vulnerabilities (Rastogi, Gloria, & Hendler, 2015). Due to the stringent security procedures at both the consumer and provider ends, there might be conflicts and leaders must resolve the disputes early before the operationalization of the solutions and the services.
- System Support: Systems control points and ownership to be identified to map with system support structure. The system administration in such kind of linkages is dynamics as there is lack of complete control of system infrastructure. Thus, consumer and provider involvement in driving the system infrastructure must be explored and agreed upon. A provider who is pro-active and extends its help and support in running the system infrastructure is considered to be more apt as a good choice.
- Service response time, throughput, and availability: Business operations should not suffer due to the poor response time, poor throughput, and unavailability of the cloud service and solution (Upadhyay, Deshpande, & Agarwal, 2009; Wu, Garg, & Buyya, 2011). The provider should be rated and opted for providing services having a minimum response time, capable of processing maximum jobs in a given unit of time and extensive accessibility and presence of service for its users.

- System utilization, scalability, and elasticity: A measure of provider's service-ability of resources for use, scalability, and elasticity of service needs to be considered (Hasan et al., 2014; Kitchin & McArdle, 2016). Understand the business cases tackled and solved by the provider in bringing optimum usage of system resources in keeping up the services, balancing the loads during high peak time and coverage of service requests distributed at varied requests.
- Reliability: Ensure that the provider offers the reliability of the cloud solution and services (Khan, Pervez, & Ghafoor, 2014; Trung & Thang, 2009; Upadhyay, 2008). Identify the strategy that is used by cloud providers to create multiple copies of the application components in delivering the services.

The concerned stakeholders must be exposed to the pain points and needs of the business units. It should be followed by the success stories of the business use cases illustrating the usage of cloud adoption addressing the needs and eliminating the pain points.

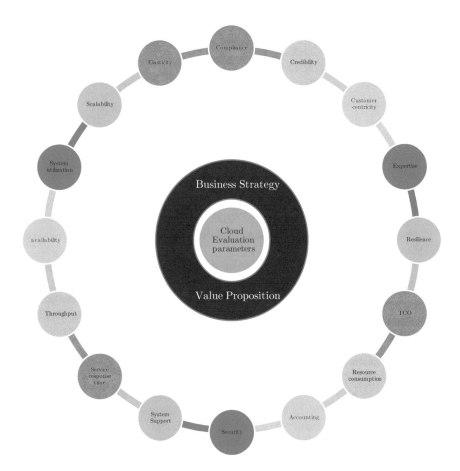

Fig. 5.1 Cloud evaluation parameters

Big Data Solutions (BDS) Evaluation

The organization should focus on exploring the big data technologies, services and infrastructures to support ingesting, sourcing, stream processing, storage and compute platforms and visualization. The technology is evolving at a breakneck pace and is very challenging for the decision makers to narrow down their focus to embrace the big data solutions. It is an excellent strategy to have a broad but profound list of parameters needed to evaluate the solutions. Before starting the big data projects, leaders have to answer few questions:

- What all data we have in place?
- What all we want to do with the data?
- What solutions can integrate data with decisions?
- How data is stored, managed, and processed?

The big data solutions offerings need to articulate the power and functionality to serve the requirements clearly. It is necessary for the organization to consider BDS vendor who is ready to address your challenges with the data use cases. Data can be stored, managed, and processed in the cloud and for the purpose you can perform the cloud solutions evaluation in addition to BDS evaluation. However, to avoid any unseen risks, it is better to review the following checklist:

- Does the BDS able to manage data with mobility and structure?
  The data has two dimensions: mobility and structure. The organization involved in harnessing the value of big data can exploit both the dimensions. For example, a company involved in researching on the technology trends can utilize batch processing (mobility dimension: static) of the unstructured data (structure dimension: unstructured). While another company can utilize stream of data (mobility dimension: continuous/stream) to control the autonomous car (structure dimension: unstructured). Understand the vendor BDS solutions to manage both the mobility and structure dimension of your business use case data.
- How effectively is the BDS capable of processing in a distributed manner and achieve parallelization?
  The secret is to process data swiftly to distribute the data and achieve parallel processing. Hadoop, as a defacto standard, provide distributed solutions. The vendor should able to articulate the strategy behind their BDS solutions to attain distributed computing and parallelization. The scalability of the data solutions is essential to store, compute, and manage data and processes.
- How does the BDS integrate with legacy and frontier technology solutions?
  In the technology landscape, you might have a constraint to work with the legacy systems but also have the flexibility to embrace the frontier technology. For such an environment, your vendor's BDS must be able to support the vertical and horizontal integration of current and future requirements. It is better to ask the vendor to provide sufficient demos to verify and validate the feasibility of the solutions.

- Does a solution provider believe in ecosystem strategy?
  A valued ecosystem strategy benefits not only the organization but also the collaborators, partners, and associates. Business leaders have to co-develop ecosystem strategy with the BDS vendor and find out in what manner the BDS vendor can be involved in solving the business use cases and extend current product and service portfolio and business models.
- How a solution provider extends the partnership relationship in more meaningful ways?
  Service effectiveness index is a critical aspect to strengthen the vendor relationship with the consumer. You would not like to have a contract with the BDS vendor who does not provide better and effective services. A BDS vendor who is flexible, pro-active and includes pre- and post-sales services (training, troubleshooting, etc.) is valued.
- How effective BDS does support exploration, visualization, and consumption of the data sets?
  Business leaders are inclined toward visualization as it is a significant way to understand the data. You might be having a particular form of visualizing and creating the narrative stories. You need to check the tools and solutions that your vendor is promising to provide to you. Look at the usability, effectiveness, and responsiveness to support your requirements.
- Does the solution intersect horizontally across business or also available for the vertical business unit's purpose?
  Some solutions need to address the common problem, but other solutions are more specific and are applied to only particular scenarios or domains. You must enquire the vendor about different solutions available for generic and specific applications. All the functionalities and features should be rated based on the quality of service (QoS) parameters in align with the business use cases. You need to identify the QoS requirement for your data use case. Few QoS parameters to consider are performance, efficiency, accuracy, reliability, maintainability, serviceability, usability, integrability, reusability. Further, you need to identify the data policy, data audit and data privacy and security policy of the vendors. Also, it is better to have a clear understanding of the administrative roles and responsibilities for the data and associated BDS solutions.

Once you have the answers to the questions, you will be in a clear position to undertake the BDS evaluation journey. You also need to perform the prioritization of the big data business use cases and identification of the value proposition. The five key areas—Source, Storage, ETL, Process, and Visualization investigation step model is helpful for the organization to evaluate the BDS vendors, Fig. 5.2.

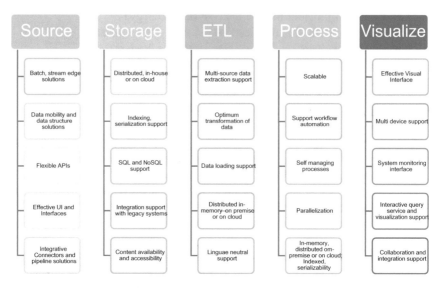

Fig. 5.2 Big data solution support review parameters

Analytical Model Evaluation

It is no value to the organization to have lots of data but no analytical support or solutions to perform insightful and actionable analytics. Analytics provide a competitive advantage to the organizations. Davenport in his seminal book "Competing on Analytics: The new science of winning" has explored the potential of analytics for organizations to compete and attain success utilizing analytics. Not all organizations use all sort of analytical solutions. Your business must drive the solutions uses cases—both current and future. When you search for the right vendor for the suitable analytics solutions, you must review the following checklist, Fig. 5.3:

- Which Analytics processes are practiced in the organization?

The Analytics solutions can be categorized as—descriptive, diagnostic, predictive, prescriptive, preventive, and autonomous solutions. Descriptive solutions provide the ability to report and visualize organizations historical and current data, and target to answer "*what happened in the past.*" The diagnostics solutions deal with the specialized domain of identifying and visualizing the cause analysis of a particular problem. At this level, the organization focuses more on domain-specific "diagnosis" of the problem space and utilizes current and historical data. The predictive solutions support to predict the future based on the past data. Once the prediction is possible about the future based on the probability or likelihood, then the prescriptive and preventive solutions can provide support in prescribing the optimum solutions and to prevent the execution of unlikely events. The autonomous solutions are the one where human involvement and support are minimal, and

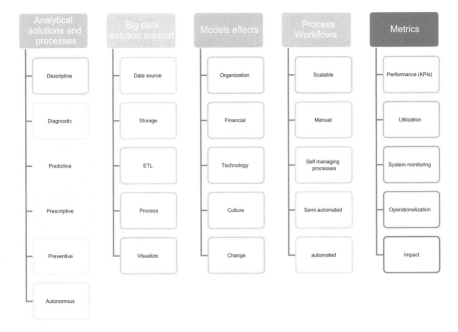

Fig. 5.3 Analytic solution review parameters

processes are cognitively advanced based on advanced machine learning techniques. You need to identify the business use cases and the category of solutions that are practiced in the organization. Once you get the vendors offerings, you need to map your business use case with the vendor's analytical solutions based on your needs.

- What big data solution support is available to perform analytics?

Once you decide on the analytical solutions necessary for the business use cases, then you must identify the big data and cloud solutions available to deploy the analytical solutions. You can refer to the cloud and big data solution section for detailed discussion on the evaluation consideration.

- What effect do analytic models have in the organization?

The effect can be organizational, financial, and technological. You need to prioritize the areas where you would like to involve your analytical vendor for the maximum benefit. Analytical solution vendor may utilize different analytical models for the same problem. For example, to identify the decision, a random forest or a simple logistic regression decision tree can be utilized, but the impact on the cost and resource utilization is enormous in the Random forest as compared to the other model. A vendor can also contribute to the success of the organizational change by providing mentorship, training, and onsite/offsite support.

- What workflows are manual, semi-automated, and automated?

You need to categorize the processes based on the choices of manual and autonomous control. Accordingly, you can identify the vendor analytical solution support for each of the processes. A frontline operator at the retail checkout counter needs to know the customer's bandwidth in buying the basket of items and appropriately in just-in-time can consider the decision of offering discounts or voucher coupons to the customer.

- What performance and utilization metrics are utilized for the analytical solutions?

The metrics are important as they quantify the usage, and outcome of the analytical solutions. As you start using the analytical solutions, you would like to measure its usage pattern, effectiveness in the operationalization, orchestration with the solution infrastructure and impact on the business strategy. Does your chosen analytical solution provide metrics to measure and quantify the meaningful attributes? Ask you, vendor, to give a demonstration of the metrics or share the testimonials of the clients that verify and validate the performance of the solutions. The vendor who can provide support to manage a sustainable change in your organization to adopt the analytically driven culture would be a good choice.

The CAB evaluation should not be done in isolation or silos, but it is an inclusive exercise and best results can be obtained when followed a collaborative, communicative, and coordinated efforts of business, IT and other key stakeholders in driving the activity, Fig. 5.4.

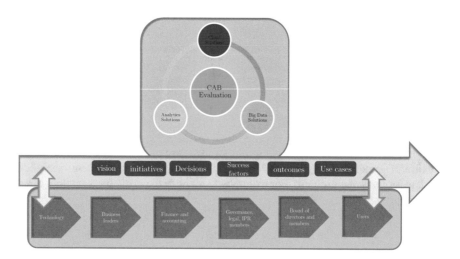

Fig. 5.4 CAB evaluation

Concluding Remarks

The CAB adoption or sourcing, though provide several benefits and opportunity to the organization, is susceptible to risks, failures, and setbacks when the adoption is materialized without considering the formal evaluation goal and exercise. A formal and structured approach is desirable for an organization to undertake such a valuable exercise. Various guidelines, criteria, and indicators necessary to evaluate the CAB solutions have been covered in great detail.

References

Beloglazov, A., & Buyya, R. (2010). Energy efficient allocation of virtual machines in cloud data centers. In *CCGrid 2010—10th IEEE/ACM International Conference on Cluster, Cloud, and Grid Computing* (pp. 577–578). https://doi.org/10.1109/ccgrid.2010.45.

Buyya, R., Garg, S. K., & Calheiros, R. N. (2011). SLA-oriented resource provisioning for cloud computing: Challenges, architecture, and solutions. In *Proceedings—2011 International Conference on Cloud and Service Computing, CSC 2011* (Figure 1) (pp. 1–10). https://doi.org/10.1109/CSC.2011.6138522.

Ding, S., Wang, Z., Wu, D., & Olson, D. L. (2017). Utilizing customer satisfaction in ranking prediction for personalized cloud service selection. *Decision Support Systems*, *93*, 1–10. https://doi.org/10.1016/j.dss.2016.09.001.

Griffin, A., & Hauser, J. R. (1993). The voice of the customer, Marketing Science, *12*(1), 1–27. https://doi.org/10.1287/mksc.12.1.1.

Gubbi, J., Buyya, R., & Marusic, S. (n.d.). *Internet of Things (IoT): A Vision, Architectural Elements, and Future Directions*, (1), 1–19. https://doi.org/10.1016/j.future.2013.01.010.

Hahn, E. D., & Knott, C. L. (2008). Assessing quality improvement initiatives when expert judgements are uncertain. *Journal of the Operational Research Society*, *59*(2), 252–258. https://doi.org/10.1057/palgrave.jors.2602484.

Hasan, S., Shamsuddin, S. M., & Lopes, N. (2014). Machine learning big data framework and analytics for big data problems. *International Journal of Advances in Soft Computing and Its Applications*, *6*(2).

Jula, A., Sundararajan, E., & Othman, Z. (2014). Cloud computing service composition: A systematic literature review. *Expert Systems with Applications*, *41*(8), 3809–3824. https://doi.org/10.1016/j.eswa.2013.12.017.

Kantarci, B., & Mouftah, H. T. (2014). Trustworthy sensing for public safety in cloud-centric internet of things. *Internet of Things Journal, IEEE*, *1*(4), 360–368. https://doi.org/10.1109/JIOT.2014.2337886.

Khan, Z., Pervez, Z., & Ghafoor, A. (2014). Towards cloud based smart cities data security and privacy management. In *Proceedings—2014 IEEE/ACM 7th International Conference on Utility and Cloud Computing, UCC 2014* (December) (pp. 806–811). https://doi.org/10.1109/UCC.2014.131.

Kitchin, R., & McArdle, G. (2016). What makes big data, big data? Exploring the ontological characteristics of 26 datasets. *Big Data & Society*, *3*(1), 1–10. https://doi.org/10.1177/2053951716631130.

Martens, B., Walterbusch, M., & Teuteberg, F. (2012). Costing of cloud computing services: A total cost of ownership approach. In *2012 45th Hawaii International Conference on System Sciences* (pp. 1563–1572). https://doi.org/10.1109/HICSS.2012.186.

Paul, S., Pan, J., & Jain, R. (2011). Architectures for the future networks and the next generation Internet: A survey. *Computer Communications, 34*(1), 2–42. https://doi.org/10.1016/j.comcom.2010.08.001.

Rastogi, N., Gloria, M. J. K., & Hendler, J. (2015). Security and privacy of performing data analytics in the cloud. *Journal of Information Policy, 5*(2015), 129–154. https://doi.org/10.5325/jinfopoli.5.2015.0129.

Rubinstein, I. S. (2013). Big data: The end of privacy or a new beginning? *International Data Privacy Law, 3*(2), 74–87. http://dx.doi.org/10.1093/idpl/ips036.

Straub, T., Kohler, M., Hottum, P., Arrass, V., & Welter, D. (2013). *Customer Integration in Service Innovation: An Exploratory Study, 8*(3), 25–33.

Teradata Corporation. (2014). Netflix: Using big data to drive big engagement unlocking the power of analytics to drive content and consumer insight non-stop testing and learning yield business results, 4. Retrieved from google.com.

Trung, P. T., & Thang, H. Q. (2009). Building the reliability prediction model of component-based software architectures.

Upadhyay, N. (2008). Structural modelling and analysis of object oriented systems: a graph theoretic system approach. *International Journal of Systems, Control and Communications, 1* (2). https://doi.org/10.1504/IJSCC.2008.021125.

Upadhyay, N., Deshpande, B. M., & Agarwal, V. P. (2009). MACBSS: Modeling and analysis of component based software system. In *2009 WRI World Congress on Computer Science and Information Engineering, CSIE 2009* (Vol. 7). https://doi.org/10.1109/CSIE.2009.964.

Wang, S., Zhang, L., Wang, S., & Qiu, X. (2010). A cloud-based trust model for evaluating quality of web services. *Journal of Computer Science and Technology, 25*(2007), 1130–1142. https://doi.org/10.1007/s11390-010-1090-7.

Wu, L., Garg, S. K., & Buyya, R. (2011). SLA-based resource allocation for software as a service provider (SaaS) in cloud computing environments. *2011 11th IEEE/ACM International Symposium on Cluster, Cloud and Grid Computing*, 195–204. https://doi.org/10.1109/CCGrid.2011.51.

Zardari, S. (2014). *Cloud Adoption: Prioritizing Obstacles and Obstacles Resolution Tactics Using AHP.*

Chapter 6
CABfication—The Explanation

It's simpler to believe in a miracle.
—William Golding, The Spire

Information about our property, our professions, our
purchases, our finances, and our medical history does not tell
the whole story. We are more than the bits of data we give off as
we got about our lives.
—David Solove
The Digital Person: Technology and Privacy in the Information
Age.

Introduction

The three important elements of the technology landscape, contributing to the formation of a Trio wave, have surpassed the business value that one can generate. No matter whether an organization is a big multinational company like Walmart having one of the powerful IT infrastructure or a start-up like Qubole, a cloud-native big data activation platform, utilize the Trio wave to exploit the opportunities. Managing the digital assets, processes, workflows, resources, infrastructure by keeping appropriate metrics and measurement enable an organization to monetize and generate value (Russom, 2011; Wang, Gunasekaran, Ngai, & Papadopoulos, 2016).

CABfication is the act of monetization and generation of business value by performing management and measurements of CAB adoption or sourcing, Fig. 6.1. Managing CAB adoption or sourcing means the establishment of directives, guidelines, policies, compliance, standard operating procedures (SOPs), governance, service-level agreements (SLAs), service-level management (SLM), roles/responsibilities, frameworks, etc. To measure the CAB adoption or sourcing success, the organization needs to develop, deploy, and execute key performance indicators (KPIs), key agility indicators (KGIs), key governance indicators (KGIs), understand value and economics (Bentes, Carneiro, da Silva, & Kimura, 2012; Kendall, 2014; Tzeng & Chang, 2011; Vecchiola, Pandey, & Buyya, 2010). Finally,

© Springer Nature Singapore Pte Ltd. 2018
N. Upadhyay, *CABology: Value of Cloud, Analytics and Big Data Trio Wave*,
https://doi.org/10.1007/978-981-10-8675-5_6

to monetize the CAB adoption or sourcing organization must focus on prioritization of strategies, initiatives, business use cases, and change and transformation execution.

There are certain barriers and challenges that an organization face to achieve CABfication. More interestingly, the true CABfication of the organization helps it to achieve the highest level of CAB maturity, as at that level it get transformed and metamorphosed into new business models, would have entered new markets and built-up new product and service portfolio. Every organization would like to achieve the highest level of the CAB maturity. Most of the organizations are at the beginning of the CAB maturity level and few have attained the higher levels. But all are apprehensive about the CAB adoption or sourcing not because they are hesitant in taking the risks or do not have resources and financials, but they are unaware of the monetization and business value that it can provide.

To understand the know-how of achieving the monetization and business value, three elements—monetize, manage, and measure process—are described. It represents the outermost layer of the CABology.

Manage

You need to establish shared understanding and responsibility ecosystems across your whole value chain. A business unit could be impacted by the insource/instream and outsource/outstream elements of the control, information, process, and artifacts

Fig. 6.1 3 M's model

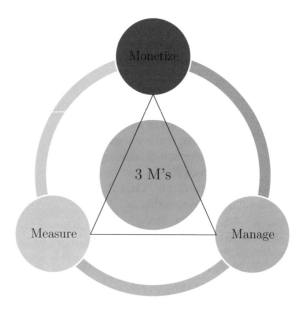

flow (Patel, Ranabahu, & Sheth, 2009). To mitigate any risks in attempting to establish a robust CAB ecosystem, an important step is to review the checklist:

- How to manage operational processes?
- How to manage cross-collaboration and joint responsibility?
- How to manage organizational change and transition?
- How to manage workload migration?
- How to manage SLMs, SLAs, and ISMs?
- How to manage risks?
- How to manage control, access, authority, and accountability?
- How to manage ownership?

The performance of business operation typically hinders due to the lack of the cross-collaboration and joint responsibility of decision making between the business team and IT (Upadhyay, 2016). Business team since many decades is considered as the forefront of the business while IT team takes a backseat and treated as the support system. General Electronics, a 140-year-old company, now foraying its journey to get transformed from the industrial and manufacturing company to industrial Internet manufacturing by involving machines, sensors and the Internet. A lot of technology landscaping is necessary to achieve such a challenging vision. At GE, IT team is not a team that takes a back seat but it goes hand-in-hand with the business team. Both the teams are responsible for the cross-collaboration and joint decision making. For example, business people at GE with their experience of the business domain knowledge, and engineering and IT team with know-how of the machine design and IT infrastructure could now know to optimize the wind machines, by letting the wind turbines to connect with the peer machines and the network and share its health to perform autonomous and informed decision. One strategy to achieve the cross-collaboration and joint decision making responsibility is to make people believe that they are inclusive and are very much part of the decision making. Also, the role and responsibility reversal of the stakeholders help them to understand elements, constraints, challenges, and opportunities out of their specific domain. IT leadership should incorporate strategy and business leadership. For example, the IT team can be considered as valuable asset in deciding the marketing campaign and delivery channels considering the wide spectrum of IT landscaping coverage. Engage your business leaders to incorporate IT strategy. For example, business people should consider secure strategies for partnership and collaboration. Security is treated as the IT strategy but business people must also consider IT strategy in business decisions. GE has decided to move its more than 9000 workloads to the cloud and to reduce its data centers footprint. Business people have incorporated security IT strategy in their decision, and they agreed on the migration of those workloads to the cloud which do not compromise their secret and confidential processes, data, and workloads.

Effective change management and transition help organization to control and avoid risks of rejection of CAB adoption or sourcing (Black, 2004). Do not worry!

Following checklist help you start managing change and transition in your organization, Fig. 6.2.

- Identify the status-quo
- Identify the destination status—where you want your organization to be. For example, if you are at "level 1" in the CAB maturity model then your status-quo is "level 1," but if you would like your company to be in "level 3" in next 2 years then your destination is "level 3"
- Identify the change and transition parameters—cultural, competency, roles, responsibilities, job profile, ecosystem
- Identify the strategic, tactical, and execution action plans
- Monitor, measure, control, and optimize the execution

Prepare your organization to perform short and long audit cycle to enforce, verify, and validate compliance, transparency, accountability, responsibility, ownership, performance, and quality. To avoid any risks in the CAB adoption or sourcing, it is necessary for the organization to build stringent service-level agreements (SLAs), service-level management (SLM), and integrated service management (ISM) system and contracts. Develop the CAB classification policy to classify the levels of the:

- analytical models to be deployed and executed
- data and security, privacy and safety controls
- Cloud services and deployment models

The CAB classification policy is important for the CABfication. For example, consider that you would like to perform execution of predictive model on the enterprise confidential data. So, instead of choosing the public cloud, the process must be executed on the private cloud. In case it is avoidable to use the private cloud for such kind of processing, then it is advised to camouflage and encrypt the data before it is put up on the public cloud.

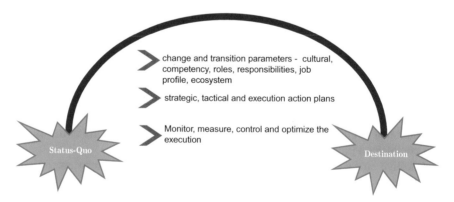

Fig. 6.2 Change and transition

In the entry point, the organization and CAB solution provider should decide and negotiate on the legal and compliance terms and conditions, Fig. 6.3. It should include:

- Detailed security SLA's address integrity
- Ownership
- Confidentiality
- Transparency
- Accessibility
- Single tenancy and multitenancy
- Ports and protocols such as security socket layer (SSL), transport layer security (TLS)
- QoS parameters
- Intellectual property ownership
- Escalations remedy and rectification timeline and accountability
- Compliance with regulation act, for example, Federation Information Security Management Act (FISMA) and Federation Risk and Authorization Program (FedRamp)

At the exit point, the CAB solution provider should agree on the following:

- Workload, process, and models [metadata] to be exported to the destination and permanently destroyed from the CAB solution provider's system
- Data should not be locked-in and to be exported to the destination and permanently destroyed from the CAB solution provider's system

Fig. 6.3 SLA consideration

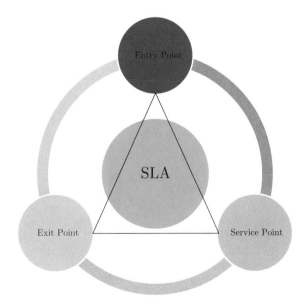

In the service point, the focus is on the:

- Verification and validation of controls
- Auditing (scheduled and on-demand) to identify any leakages, damages and health of the organization assets, resources and infrastructure due to CAB solutions.

The health of the SLA is managed and monitored by the service-level management system. It is advised to deploy a robust SLM to avoid any failures in SLA attainment. Following points need to be considered, Fig. 6.4:

- Ensure SLM to monitor and notify the status and its deviation from the expected level to the concerned stakeholders
- Prepare SLM to invoke performance improvement plans during the failure of SLA
- Deploy SLM to perform continuous comparison of SLA with the agreed, expected and actual parameters

Further, it is advisable to adopt an integrated service agreement (ISM) with the CAB solution provider to achieve business and operational excellence and to mitigate any risks. It provides internal and external CAB service visibility to the customers. Few considerations for the CABfication are:

- The accountability and the onus of the service to your customers is on you, even you utilize CAB solution provider in your value chain.
- Jurisdiction mandate to hold data, IPR, applications; some countries promote and put constraints on building the CAB solution by the country specific citizens only, more specifically cloud solutions. Thus, thorough study to be done before adopting the CAB solutions.

Fig. 6.4 SLM requirement

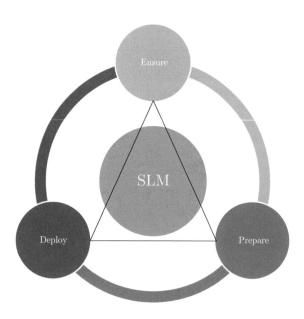

Measure

One of the fundamental principles behind monetization is "to measure." You cannot manage if you cannot measure. When you measure aspects of the business, then you have an opportunity to mitigate any risks, initiate performance improvement plans, take informed and actionable decisions, enter new markets and metamorphosed to new business models (Giessmann & Stanoevska-Slabeva, 2012). Few artifacts are important to consider to measure in the organizational framework for the CAB adoption:

- Incidents
- Knowledge
- Workforce performance
- Provisioning and release
- Assets
- Configuration
- Service request, delivery, and performance
- Change
- Value
- Economics
- Key performance indicators
- Key agility indicators
- Key governance indicators
- Quality of Service
- Analytical Models performance

Once should use the standardized measurement framework to measure for the benefit of the tracking, monitoring, controlling, managing, and monetization.

Monetize

The organization must tap on the direct and unintended consequence of the CAB adoption. Mostly, at the initial stage the business leaders concentrate their efforts in streamlining intended interests and business use cases leaving aside the second- and third-order consequences. For example, Citibank deployed their first ATM to optimize their workforce and that was the first-level intended consequence. But, the solution helped Citibank to gather much more and deep transactional level data of the customers. They could identify when and where customers are withdrawing money, so that they can strategize the service offerings and placement of their ATMs. Now, that was the second-order consequence of Citibank's action of rolling out the ATM. They monetize at both the levels—workforce optimization and service offerings.

Organizations need to understand few principles behind monetization of CAB adoption:

- Exploiting Big data: Business units should come together to share the common interest regarding the big data utility. It might be possible that all sorts of data might not be useful at the first-level order and intended consequence for the business use case. It is necessary to stress upon the developed of repository covering second-, third-, and higher-order intended and unintended consequences. Few checklist questions are important to generate ideas:

  – What characteristics are important for the business use case?
  – What are the priority of business use case in terms of first-, second-, and higher-order consequences?
  – What big data can be used across business units?
  – What category of stakeholders gets affected by the big data?

- Exploiting Analytical models: One should explore the different ways to approach decision making for the better and faster decisions. Insurance companies in the USA perform behavioral analytics on the taxi driver data. Each taxi driver gets a health scorecard depending on the way that driver drives the taxi. The health insurance policy is costly, and thus, a particular premium deduction is offered to the taxi driver for the health policy. The intended consequence and the primary business use case of the insurance company are to provide viable insurance plan to the customers. But the higher-order consequence caused the improvement in the ethical and behavioral aspects of a person. Organizations need to deploy and experiment models to test, verify, and validate the decisions. Try out different analytical models such as—descriptive, diagnostic, predictive, prescriptive, preventive, and autonomous to check the performance of the solution. Few checklist questions are important to generate ideas:

  - Which segment of customer to target?
  - Why failure of service happened?
  - What will be the optimum workforce required to deliver 10,000 products in the quarter-1?
  - Which product customer will buy next?
  - What controls are required to get the optimum balance between performance and cost?
  - Which patient needs cardiac care between 4 and 12 h?
  - When to schedule the maintenance activity of jet engine?

- Exploiting cloud services and models: Identifying the appropriate cloud service and deployment models, help organization to monetize effectively. GE's oil and gas business unit optimized its total cost of ownership (TCO) by adopting cloud for its services. Netflix developed from the scratch all its resources, processes, and workflows for the optimized use of cloud service to provide unmatched viewing experience to its viewers. While migrating to cloud, Netflix could migrate all its workflows and application "as-is," but to avoid replicating the problem and limitation of data centers they became cloud-native. Different service models—SaaS, PaaS, IaaS, and XaaS—are available that could be deployed over private, public, hybrid, or community clouds. Few checklist questions are important to generate ideas:

- How much compute and storage capacity is required?
- What are the core, critical, moderate and easy processes, workflows, and applications required for the operational purpose?
- What data is critical?
- What is customer's peak time to avail service?
- How much infrastructure load is witnessed during product and service accessibility?
- What is the total cost of ownership of the IT infrastructure?

Concluding Remarks

Managing the digital assets, processes, workflows, resources, infrastructure by keeping appropriate metrics and measurement enables an organization to monetize and generate value. Organizations are encouraged to use CABfication which is the act of monetization and generation of business value by performing management and measurements of CAB adoption or sourcing. There are certain barriers and challenges that an organization face to achieve CABfication. More interestingly, the true CABfication of the organization helps it to achieve the highest level of CAB maturity, as at that level it gets transformed and metamorphosed into new business models, would have entered new markets and built-up new product and service portfolio. To understand the know-how of achieving the monetization and business value, three elements—monetize, manage and measure process—are described.

References

Bentes, A. V., Carneiro, J., da Silva, J. F., & Kimura, H. (2012). Multidimensional assessment of organizational performance: Integrating BSC and AHP. *Journal of Business Research, 65*(12), 1790–1799. https://doi.org/10.1016/j.jbusres.2011.10.039.

Black, T. (2004). *Philosophy of mind and metaphysics. Lecture II: Identity over time and over change of composition.* Lecture Notes California State University, Northridge. 1–6.

Giessmann, A., & Stanoevska-Slabeva, K. (2012). Business models of platform as a service (PaaS) providers: Current state and future directions. *Journal of Information Technology Theory and Application, 13*(4), 31–55.

Kendall, K. (2014). Social responsibility and performance excellence, (April).

Patel, P., Ranabahu, A., & Sheth, A. (2009). Service level agreement in cloud computing. *Kno.e. Sis Publications, 14*(6), 126–131. https://doi.org/10.1007/978-1-4614-1614-2.

Russom, P. (2011). Big data analytics. *TDWI Best Practices Report,* 1–35. https://doi.org/10.1109/ICCICT.2012.6398180.

Tzeng, G., & Chang, H. (2011). Applying importance-performance analysis as a service quality measure in food service industry. *Journal of Technology Management & Innovation, 6*(3).

Upadhyay, N. (2016). SDMF: Systematic decision-making framework for evaluation of software architecture. In *Procedia computer science* (Vol. 91). https://doi.org/10.1016/j.procs.2016.07.151.

Vecchiola, C., Pandey, S., & Buyya, R. (2010). High-performance cloud computing: A view of scientific applications. In *Proceedings of 10th International Symposium on Pervasive Systems, Algorithms, and Networks*. https://doi.org/10.1109/I-SPAN.2009.150.

Wang, G., Gunasekaran, A., Ngai, E. W. T., & Papadopoulos, T. (2016). Big data analytics in logistics and supply chain management: Certain investigations for research and applications. *International Journal of Production Economics, 176,* 98–110. https://doi.org/10.1016/j.ijpe.2016.03.014.

Chapter 7
CAB Proposition—The Way Forward

Because if you take a risk, you just might find what you're looking for.

—Susane Colasanti, Take Me There

It is always important to know when something has reached its end. Closing circles, shutting doors, finishing chapters, it doesn't matter what we call it; what matters is to leave in the past those moments in life that are over.

—Paulo Coelho, The Zahir

Introduction

In the report presented by HCL technologies covering IDC predictions, it is mentioned that the predicted CAB worldwide market size is around $117 billion and monetization capability is of $400 billion by 2020. The CAB revolution is not coming, but it is underway, and businesses need to prepare themselves to take the best of it to manage, control, operate, and create value. GE has a target to achieve $15 billion sales by 2020 and willing to spend $1 billion to develop business value out of utilizing CAB to leverage data from sensors—gas turbines, machines, jet engines, oil turbines, edge technology, etc (Winig, 2016). Businesses can lose the opportunities and benefits unless their efforts are focused in the areas that matter the most. Organizations face challenges to understand *"how CAB can power their business initiatives."* The inclusion of CAB value proposition is necessary to understand the impacts of the CAB across your enterprise network. The business users need to ask tough and challenging questions keeping in the account of the changing technology ecosystem, customer behavior, and market dynamics contributing to following assertions:

- Your CAB adoption becomes a business core and cost efficient
- Your organization achieves operational excellence and increased agility
- Your stakeholder's concern are considered valued, and they feel inclusive
- Your CAB projects are conceived, managed, and executed enterprise-wide

© Springer Nature Singapore Pte Ltd. 2018
N. Upadhyay, *CABology: Value of Cloud, Analytics and Big Data Trio Wave*,
https://doi.org/10.1007/978-981-10-8675-5_7

- Your organization manifested to performance and profit-driven engine
- Your organization embraces new ventures and portfolio of services and products.

Business questions (select few)

Q.1. Who are my customers and more importantly valuable customers?
Q.2. What channels my customer use to utilize my services and products?
Q.3. What products and services are contributing to my customer's satisfaction?
Q.4. What products and services are responsible for my customer's dissatisfaction?
Q.5. What are my successful and engaging campaigns?
Q.6. What are the barriers to the adoption of technology?
Q.7. What are the challenges in the offerings of products and services?

Business users find the questions above simple but then why these questions are not being answered profoundly and precisely. These questions are in the businesses for several decades, and with the help of CAB framework they can rethink the issues, get the viable perspective, and able to get the feasible solutions.

Let us look at each of these questions thoughtfully from the perspective of CAB framework adoption. The CAB has redefined and quantified the terms such as valuable, success, important, and goodness.

Who are my customers and more importantly valuable customers? Perhaps the questions are directed toward the identification of the customers, and I believe that none of the businesses will excel until they identify the customer base for their services and products. The CAB framework will help in understanding not only about who all are the customers but also who all are most influential, engaging, supportive, profitable, and trustworthy. The organization further can identify the correlation between profitable and influential customers and target base the campaign and recommendations. It may happen that the most profitable customers might not turn out to be most valuable customers due to the negligent net influential or advocacy effect on the market.

What channels my customer use to utilize my services and products? The customer can use multimode channels to utilize services and products and thus leave its [digital] footprints. The organization can use these [digital] footprints for addressing challenges related to using of services and products on the multimode channels. A customer might be interested in immersive technology to use certain types of products, and thus organization can grasp the customer journey with the help of [big] data and analytical processes.

What products and services are contributing to my customer's satisfaction? Customer satisfaction index is a most sought after KPI for an organization to gauge the level of customer satisfaction. The organization by using CAB framework is not only capable of getting the just-in-time customers data but also to pump that data into customer's analytic engine to understand *"what is contributing to customer's satisfaction."* An Amazon's kindle get continuous data of its customer base to understand—*"the taste of customer's reading"* and accordingly able to customize its offerings, recommendations, and suggestions.

What products and services are responsible for my customer's dissatisfaction? The success of product and service offerings are not just be based on the good reviews and customer feedback, but by incorporating customer feedback and reviews in a continuous and timely fashion in improvising the quality of goods and services. The CAB framework can be utilized to converge the opinion, sentiments, liking, and interests of customers to identify the improvement points. It also helps in segregating the fake customers from the trust worthy customers. The CAB can act as a whistle-blower to identify the product and service that can be a potential barrier to company's success.

What are my successful and engaging campaigns? Billboards, smart audio, video, gaming, and social media interfaces are some examples to capture the behaviors and engagement level of the customers or potential customers. Some campaigns work better than the other campaigns, few works better regarding particular channel and CAB provides a way to understand the difference between successful and engaging campaign. Are all engaging campaigns are successful? Perhaps the answer is hidden inside the meaning of "success" to you. If you think that more traffic to your campaign is the "success" parameter and you find that eventually it is materialized, then you have attained success. But, if your "success" means conversion rate to monetize campaigns, then it might happen more traffic did not turn out to be a monetizing activity and resulted into a failure. In certain cases, some campaigns like *"organ donation"* might be engaging but not successful as people do not step out to donate organs. The [big] data and analytical processes further can help organization running *"organ donation campaign"* to identify and classify potential customers who can provide consent for the *"organ donation"* and then pump the campaign accordingly.

What are the barriers to the adoption of technology? Though technology is believed to contribute immensely to the value of the business, unfortunately, it is not being adopted swiftly across the organization. The CAB framework can be used to identify and overcome the barriers across the organization. The data repository (batch and stream) related to the employee engagement, performance, and training, business operations, productions, manufacturing, marketing, sales, etc., can be utilized to understand the barriers. The top barriers for the organization to adopt the technology are inadequate human resource training and awareness for the technology usage, inefficient communication, and collaboration tools, inappropriate culture, vision, and motivation.

What are the challenges in the offerings of products and services? Mostly, the challenges that businesses count are related to finance, infrastructure, materials, human resource, and environment. Walmart with its RetailLink system open its retail data to the entire supply chain which help suppliers and vendors to understand the health and value of their product. Several challenges related to products and services offering could be addressed due to the availability of the RetailLink. Also, questions like—*"who are my potential and valuable customer?"*, *"when and what to stock?"*, *"when and how much to discount"* were addressed and helped Walmart to attain leadership. The company could identify the valuable suppliers regarding time to delivery and good quality products and services in a predictable manner.

CAB Business Drivers

The organization must understand the value proposition to adopt CAB framework. The CAB business drivers, Fig. 7.1, are the key elements of the CAB value proposition that help the organization to charter its business value creation pathways and action plans. At the initial phase, organization needs to conduct feasibility assessment, planning of the engagement and determination of the primary, secondary, and other stakeholder's requirements, and service and product delivery approach. Further, evaluation of the business and IT ecosystem is to be followed by considering aspects of management, people, process, information, and technology (Shrestha, Cater-steel, Toleman, & Tan, 2014). To manage the organization's transformation and growth approach, capture the following items and establish the traceability of the relationships that generate the business value:

- Organization's goals
- Organization's business mission and objectives
- Organization's business functions and cross-functions
- Organization's key information and interaction value chain with partners, vendors, suppliers, customers, and external entities
- Organization's tangible and intangible assets
- Organization's business, information, and process models
- Organization's key performance, success, and agility parameters
- Organization's resource—people, material, technology

At this stage, the organization will commence its journey to exploit CAB framework.

Fig. 7.1 CAB business drivers

Cloud Drivers

Information, process, workloads, and resources that are relevant to the core of the organization are to be identified. In addition, determine the entities that can be accessed, kept or migrated on the cloud. A proper mapping of process, resource and application with the business ecosystem considering a set of software, middleware and hardware building blocks is recommended (Larai, 2015; Li, Yang, Kandula, & Zhang, 2010; Qu, Wang, Orgun, & Bouguettaya, 2014). A transition plan based on the gap analysis is to be executed. Here, you would like to consider the training and competency building exercise for the cloud enablement space. Figure 7.2 shows cloud adoption or sourcing drivers.

Determine Business Line

The entry point for the cloud enablement is to analyze whether an organization should go a phase-wise implementation of cloud adoption or the project-wise. Both have its benefits and challenges. Stakeholders can make or break the success of the cloud adoption. The key decision maker's involvement, support, and motivation toward driving the business initiative is a key driver of organizational changes, governance, partnerships, and cloud technology adoption (Benlian, Hess, & Buxmann, 2009; Christi, Oishi, Pinheiro, Vasconcellos, & Nelson, 2010; Dutta & Bose, 2015; Zardari, 2014). Also, consider evaluating usage of private, public, and hybrid clouds for business operations and service offerings. A proof of concept can be executed to get the quick feasibility analysis. Do not consider proof of concept to

Fig. 7.2 Cloud adoption or
sourcing drivers

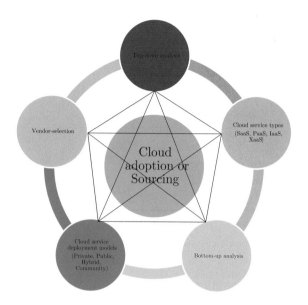

be production deliverable. Once you gain confidence with the proof of concept, you may go ahead with the pilot project. A pilot project will help the organization to understand the change needed in the configuration, control, and management of cloud services.

One way is to choose the top-down analysis where you rely on incorporating cloud investments based on your current business architecture. At this step, it is quite challenging for the organization to include the best-fit space as the organization is not sure about the cloud perspective and thus the focus is on the business processes. The top-down approach guides the business units to consider Business-as-a-Process and Software-as-a-Service solutions to support cloud and non-cloud related decisions. The primary focus here is not to assess the cloud solutions or vendors rather how these solutions can drive the business strategy and vision. Each business unit needs to identify sourcing options of the business operations, processes, and services on the cloud. The business leaders have to focus on identifying the key business competencies and performances. The business competencies further are classified under the business domain. For example, in distribution business domain, the primary competencies refer to capabilities, skillset, proficiencies, and expertise to perform its associated business operations. Some business unit operates independently while some needs orchestration with other business units. Further, the decision leader's task is to identify differentiated and non-differentiated business units and business operations processes, services, and applications. It is worthwhile to consider non-differentiated entities as easy targets for the cloud adoption. The sourcing and adoption of cloud further be considered based on the security and compliance requirements.

The other way is to perform the bottom-up analysis. The primary focus of the key decision makers is to adopt cloud based on the IT focus, and they are sure about the impact that the cloud will have on its IT infrastructure and processes. The IT unit becomes the key target to map the functional usage with the cloud offerings. The major concern when the bottom-up process to be adopted when:

- Organizations have reached their physical capacity for the networking, storage, and processing facility
- Organizations have to consider mission-critical business operations
- Organizations are running legacy systems and services
- Organizations have less IT talent to support business units

An organization "Xees" finds that its customer engagement business service is degrading, and it is prone to losing the customer base, market reputation, and the market size. The key decision holders held the meeting and found that the poor performance of the degraded performance of the customer engagement service is due to the poor in-house customer service support and system. The top-down analysis of the situations reveals that the customer management is to be considered as a business competency and to be outsourced since it falls in the gray area of non-differentiated competency. The key decision makers consider the customer's engagement to be (out)sourced to cloud as Business-as-a-Service (BPaaS) rather the traditional business process outsourcing (BPO) candidate to gain the competitive advantage. By default, the BPaaS supports the offloading of the IT-associated workloads and services.

Establish Vendor Selection Criteria

It is the critical matrix by which your organization can ensure its presence, power, and leadership in utilizing and providing the cloud services. Care should be taken to avoid any vendor lock-in feature for the data and services (Ghazouani & Slimani, 2017). It is a necessary exercise as you might plan later on to host your applications and services in-house, and then you want to avoid any challenges in migrating those applications and services which might occur due to cloud vendor propriety [meta] data formats. It is always better to form a steering committee comprising of IT and business leaders. A committee having balanced mix of the cross-business functional area will have an added advantage to see through the alternative options for cloud sourcing and adoption. It is always good option to opt for migration or customization of business services in the cloud. Choosing a vendor, however, is not an easy task. A local vendor might be suitable due to cultural and local adaptivity, rather than the global vendor who might have experience. A balance between risk and opportunity is to be looked carefully. It is recommended to get the client list from the vendor and also the testimonials recommending the particular service where your organization is interested in investing. Both business and IT users must be involved to understand the vendor's offerings. In few cases, a proof of concept of the vendor solutions can be deployed by involving the solution specific consultant.

Analytics Drivers

The analytics drivers help the organization to understand more about their business processes and potential values. Business leaders by operationalizing analytics can create a new set of business value dictionary comprising of verbs and adjectives such as—predict, forecast, score, recommend, prevent, and optimize; good, accurate, high performance, just-in-time, cutting edge, (Crawford & Schultz, 2014; Modarresi, 2016; Saaty, 2010). Table 7.1 describes critical questions related to analytics requirements. Figure 7.3 describe the analytics drivers.

Organizations need to build the people, management, and infrastructure competency to operationalize the analytic models.

- Identify methods to score models based on accuracy, compliance, performance, regulatory requirements
- Develop (semi)-automated process of model deployment
- Execute decisions autonomously.

Table 7.1 Analytics business questions

Analytics	Business question's
Describe	What's the benefit and potential of leveraging descriptive analytics to understand the business operations, staffing, sourcing, materials management, market campaigning
Diagnose	What's the benefit and potential of leveraging diagnostic analytics to diagnose the business operations, staffing, sourcing, materials management, market campaigning
Predict	What's the benefit and potential of leveraging predictive analytics to predict the business operations, staffing, sourcing, materials management, market campaigning
Forecast	What's the benefit and potential of leveraging forecasting analytics to forecast the business operations, staffing, sourcing, materials management, market campaigning
Score	What's the benefit and potential of leveraging predictive analytics to score the customers for fraud, retention, cross sell, up sell, likelihood to recommend
Recommend	What's the benefit and potential of leveraging predictive analytics to recommend the customers for buying, upgrading, changing, modifying service and products
Prevent	What's the benefit and potential of leveraging predictive analytics to prevent the service and product failures, fraud, crime, physical and monetary loss
Optimize	What's the benefit and potential of leveraging predictive and prescriptive analytics to prescribe and optimize the business operations, staffing, sourcing, materials management, market campaigning

Fig. 7.3 Analytics drivers

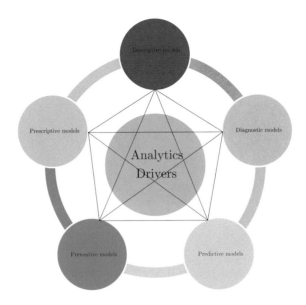

Big Data Drivers

The big data drivers are the key components of the CAB value proposition. The organization that doesn't know what data they have or could they gather and acquires is unable to generate the business value. Netflix, an entertainment giant, had focused on what data it had to produce its original content that other production houses did not. Netflix has 120 million subscribers in more than 190 countries and records every viewer "action" per day including when a viewer views particular content, pause, rewind, or fast-forward it; perform content-related searches, provide content ratings and devices utilizes to act on it.

The organization must identify the [big] data that it has or can get to utilize it to drive the business use cases. Following big data drivers are important, Fig. 7.4:

- Operational data
- Enterprise "Dark" data
- Public data
- Commercial data
- Social media data

Operational Data

Access to more detailed operational data helps business users to understand business use cases and target specific decisions in context (Demchenko, De Laat, & Membrey, 2014; Datameer, 2017; Fosso Wamba, Akter, Edwards, Chopin, &

Fig. 7.4 Big data drivers

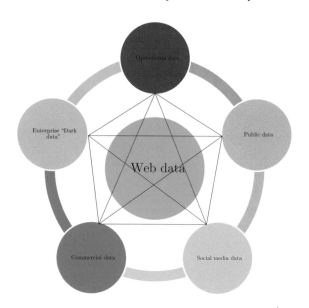

Gnanzou, 2015; Kamvysi, Gotzamani, Andronikidis, & Georgiou, 2014). Every organization has some transactional data that can be utilized for various purposes. Organization purposely or in a process gathers data via sensor or process monitoring related to, Fig. 7.5:

- customers,
- point-of-sale transaction,
- call record,
- log data,
- radio frequency identification,
- credit card transactions,
- smart meters,
- security camera feeds,
- wireless signals,
- insurance claims,
- Internet of Things (internet connected devices).

Big data technology has the potential to access, manage, and analyze all the operational data. Business leaders with the help of more detailed and granular level data are able to understand complex situations and generate insightful actions. The edge decisions are crucial as these are dynamic and based on certain configuration and situations in a context. For example, location-based service to deliver optimum service offerings and recommendations to a customer will be possible when a customer is locked-in particular location proximity, is approachable and accessible.

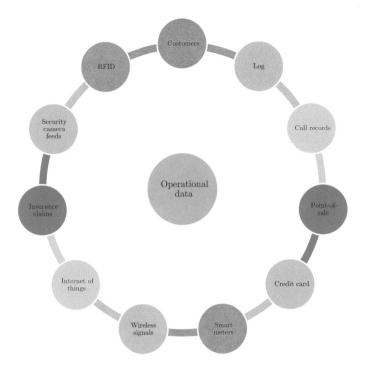

Fig. 7.5 Operational data

Such time of low-latency data is to be considered by the organization to drive their business use cases and initiatives.

Public Data

The government has conceived many projects where they have to open their data to be utilized for benefits of citizens. In India, the government has initiated several projects for open data support through the Web portal https://data.gov.in/. As per the National data sharing and Accessibility Policy—2012

> …Asset and value potentials of data are widely recognized at all levels. Data collected or developed through public investments, when made publicly available, and maintained over time, their potential value could be more utilized. There has been an increasing demand by the community, that such data collected with the deployment of public funds should be made more readily available to all, for enabling rational debate, better decision making, and use in meeting civil society needs.

[National data sharing and Accessibility Policy—2012; https://data.gov.in/; accessed January 25, 2018]

The community and government at large will benefit from public data, Fig. 7.6.

Fig. 7.6 Public data

- Maximizing usage: Organizations and community can better utilize the government data for the benefit of the public resource. For example, the retailer can use the edge delivery by tagging, linking, and utilizing the post-office location data.
- Avoiding duplication: Organizations and community can achieve cost savings as multiple collection centers for the same data can be curbed.
- Maximizing integration: The common standards for integration need to be adopted to achieve the seamless integration and worth of the integrated and aggregated data.
- Ownership information: The responsible owners can be collaborated, communicated and involved in envisioning the usage and utility of data.
- Improved decision making: Readily available public data help community and organizations to take informed decisions related to asset management, disaster management, routing optimization, development planning, improving living conditions, and national security.
- Equity of information: The better access to the information can be achieved by having the open data transfer policy.

Social Media Data

Social media help organizations to be abreast with the trends, customer opinion about their service and offerings, competitive market information, market dynamics,

 **Ministry of Railways** ✔
@RailMinIndia

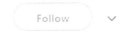

"Clean My Coach" SMS Based Coach Cleaning System Introduced In Northeast Frontier Railway.

nfr.indianrailways.gov.in/view_detail.js ...

6:03 PM - 24 Mar 2016

Fig. 7.7 Railway ministry tweet "SMS-based coach cleaning system."

potential market, etc. The politics have changed its way as the public is directly connected to the leaders and able to communicate with them. Social media data have been utilized to predict flu outbreaks, raise funds, raise awareness of climate change issues. Organization needs to incorporate the social media data into their digital policy. For example, the Ministry of Railway in India has started an SMS/ Tweet initiated "Clean my coach" service as shown in Fig. 7.7.

Anyone who is traveling by train if not satisfied with the coach cleaning can SMS/tweet to the railway ministry. The railway ministry is equipped with the necessary resource to provide the quality service to delight the passengers. Social media is not only empowering private organizations but also transforming the government bodies. In another example, a passenger traveling on a Jaipur–Pune superfast express train on June 2, 2017, found that the phone charging port is not functioning. A tweet of him to the railway ministry triggered the socket fixing activity, and the port was fixed within an hour (https://economictimes.indiatimes. com/industry/transportation/railways/travelling-in-a-train-just-tweet-your-complaint-and-get-quick-help/articleshow/58976711.cms).

Commercial Data

Organizations can also tap the commercial legal available data from the data aggregators and brokers. One can get the third-party data under the purview of regulations, and it is utmost importance to check the privacy, transfer and usage law associated with the data before investing in data acquisition.

Enterprise "Dark Data"

Organizations have a huge archive of unutilized data such as—emails communi-cations, oral history, contract documents, memos, notices, multimedia, and system logs, patents, copyrights documents, designs, and other intellectual property

Fig. 7.8 Enterprise "Dark data"

artifacts, which are untapped, Fig. 7.8. To get a competitive advantage, it is necessary to understand, process, and utilize such unstructured data. The big data technology solutions provide features and functionality to integrate and tap the unstructured data with the operational—transactional data to generate radically insightful actions.

Web Data

The Web is the biggest source of the data, and the organization which gets hold of the Web data can attain a competitive advantage. One can scrap the content from the competitor's, partners, or industry Web site to understand their business models, differentiators, and competitiveness. Competitive Analytics has developed a value-added scrapping solution that promises to provide valuable and insightful information. Figure 7.9 shows the detail aspect of the Web data.

Business Value Proposition Model

The business value proposition model, Fig. 7.10, guides organization to achieve the value proposition considering the adoption of CAB framework.

Fig. 7.9 Web data

Value Design

In this phase, the organization needs to work on the strategic and tactical elements of the CAB framework. It comprises the identification of:

- CAB vision—enterprise-wide
- Business strategy and initiatives

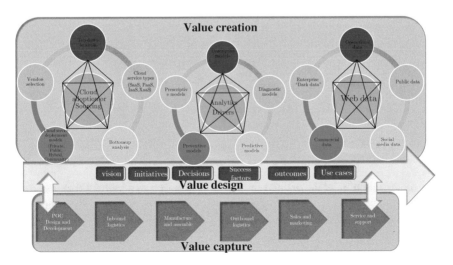

Fig. 7.10 Value proposition

- Business decisions
- Business success factors
- Business outcomes
- Business use cases

Value Capture

In this phase, the organization needs to capture the value considering its complete supply chain—upstream and downstream entities, collaborators, partners and primary, secondary, and other activities. It comprises the value capture at:

- Upstream and downstream
- POC, design, and development
- Inbound logistics
- Manufacture and assembly
- Outbound logistics
- Sales and marketing
- Service and support

Value Creation

In this phase, the organization needs to create the value by considering CAB business drivers. It comprises the value creation by:

- Adopting or sourcing services, products, and offerings to cloud.
- Developing, managing, deploying, and executing analytical models—descriptive, diagnostic, predictive, preventive, and prescriptive.
- Acquiring and gathering big data–Web data, public data, commercial data, dark data, operational data, and social media data.

Concluding Remarks

The organization must understand the value proposition to adopt CAB framework. The CAB business drivers are the key elements of the CAB value proposition that help the organization to charter its business value creation pathways and action plans. CAB business value proposition helps to manage the organization's transformation and growth approach, and establish the traceability of the relationships that generate the business value.

References

Benlian, A., Hess, T., & Buxmann, P. (2009). Drivers of SaaS-Adoption—An empirical study of different application types. *Business & Information Systems Engineering*, *1*(5), 357–369. https://doi.org/10.1007/s12599-009-0068-x.

Christi, M., Oishi, M., Pinheiro, E., Vasconcellos, De, G. & Nelson, R. (2010). The adoption of new technology: Conceptual model and application, *Journal of technology management & innovation*, *5*(4).

Crawford, K., & Schultz, J. (2014). Big data and due process—Toward a framework to redress predictive privacy harms. *BCL Rev.*, *55*, 93–128. https://doi.org/10.1525/sp.2007.54.1.23.

Datameer (2017). Paf: Using big data analytics to increase operational efficiency. https://www.datameer.com/wp-content/uploads/2017/04/Using-BigData-Analytics-Improve-Operational-Efficiency-PAF-1.pdf.

Demchenko, Y., De Laat, C., & Membrey, P. (2014). Defining architecture components of the Big Data Ecosystem. In *2014 International Conference on Collaboration Technologies and Systems, CTS 2014* (pp. 104–112). IEEE Computer Society.

Dutta, D., & Bose, I. (2015). Managing a big data project: the case of ramco cements limited. *International Journal of Production Economics*, *165*, 293–306. https://doi.org/10.1016/j.ijpe.2014.12.032.

Fosso Wamba, S., Akter, S., Edwards, A., Chopin, G., & Gnanzou, D. (2015). How "big data" can make big impact: Findings from a systematic review and a longitudinal case study. *International Journal of Production Economics, 165,* 234–246.

Ghazouani, S., & Slimani, Y. (2017). A survey on cloud service description. *Journal of Network and Computer Applications*, *91*(April), 61–74. https://doi.org/10.1016/j.jnca.2017.04.013.

Kamvysi, K., Gotzamani, K., Andronikidis, A., & Georgiou, A. C. (2014). Capturing and prioritizing students' requirements for course design by embedding Fuzzy-AHP and linear programming in QFD. *European Journal of Operational Research*, *237*(3), 1083–1094. https://doi.org/10.1016/j.ejor.2014.02.042.

Larai, M. B. (2015). A review of service selection in cloud computing, (October).

Li, A., Yang, X., Kandula, S., & Zhang, M. (2010). CloudCmp: Comparing public cloud providers. In *Proceedings of the 10th Annual Conference on Internet Measurement—IMC'10*, 1. https://doi.org/10.1145/1879141.1879143.

Modarresi, K. (2016). Recommendation system based on complete personalization. *Procedia Computer Science*, *80*, 2190–2204. https://doi.org/10.1016/j.procs.2016.05.379.

Qu, L., Wang, Y., Orgun, M. A., & Bouguettaya, A. (2014). Cloud service selection based on contextual subjective assessment and objective assessment (Extended Abstract). In *Proceedings of the 13th International Conference on Autonomous Agents and Multiagent Systems (AAMAS 2014)* (pp. 1483–1484).

Saaty, T. L. (2010). Economic forecasting with tangible and intangible criteria: The analytic hierarchy process of measurement Ekonomsko Predvi Đ Anje Primenom Merljivih I Nemerljivih Kriterijuma: Analiti Č Ki Hijerarhijski, (1), 5–46.

Shrestha, A., Cater-steel, A., Toleman, M., & Tan, W. (2014). A method to select IT service management processes for improvement. *JITTA: Journal of Information Technology Theory and Application, 15*(3), 31–56.

Winig, L. (2016). *GE's big bet on data and analytics seeking opportunities in the internet of things, GE expands into industrial analytics.* MIT Sloan Management Review.

Zardari, S. (2014). *Cloud adoption: Prioritizing Obstacles and Obstacles Resolution Tactics Using AHP.*

Chapter 8
CAB Implications—The Affairs

Whenever a theory appears to you as the only possible one, take this as a sign that you have neither understood the theory nor the problem which it was intended to solve.

—Karl R. Popper

When I first started in Formula 1, I tried to ignore the fact I was the first black guy ever to race in the sport. But, as I've got older, I've really started to appreciate the implications.

—Lewis Hamilton

Introduction

Your decision to adopt CAB framework requires carefully thought out transition and transformational planning. The leaders must envision the type of change and guide the organization to undergo such change. The mere adoption of CAB-related technologies does not help the organization to attain CAB leadership. It should be present as part of its DNA. The organizations like Amazon, Walmart, and Dell have focused their efforts on improving the value of their supply chain. Improvement in the customer service and loyalty has been the focus of Harrah's, Capital One, and Neiman Marcus efforts; some organizations like Progressive Insurance and Marriott focused on pricing; Honda, Toyota, and Intel targeted on improving product and service quality. But why the difference? The only reason to do is to be driven by the strategy to gain the business value. Such leading organizations do not just put their efforts in CAB adoption randomly but a thoughtful, and inclusive realization has helped them to enter into the CAB space. It is paramount essential to the success of any business initiative that it must support the business strategy (Chen, Shih, Shyur, & Wu, 2012; Gupta & Consulting, 2010; Upadhyay & Deshpande, 2010). Not every organization needs the same level of technology and resource space, implementation efforts, and action plans. The CAB business strategy provides a plan that helps the organization to begin its journey in the CAB space, defines its long-term goals, the way that organization plan to achieve these goals and the way that an organization plans to gain a competitive edge considering differentiator

© Springer Nature Singapore Pte Ltd. 2018
N. Upadhyay, *CABology: Value of Cloud, Analytics and Big Data Trio Wave*,
https://doi.org/10.1007/978-981-10-8675-5_8

factors and offerings. The CAB strategy sets the directions for the organization to move into the CAB adoption space. The organization needs to raise the questions to identify its marketplace, offerings, differentiation, uniqueness, opportunities, challenges, etc. (Chen, 2012; Kshetri, 2014).

- What marketplace are we in?
- What marketplace are competitive?
- What differentiators our products have?
- How do we improve and compete in the marketplace?
- What is opportunity space available?
- What are the challenges to foray into the marketplace?

For Marriott, it was pricing; for Capital One it was customer service and loyalty; for your company, it might be different. These strategic decisions set the tone for the organizations to put all its efforts in streamlining processes, and resources by considering the tactical decision on the operational basis. Tactical decisions must be aligned with the strategic decisions. To fully adopt the CAB framework, organization need to develop CAB business strategy, CAB strategic decisions, CAB tactical decisions, Fig. 8.1.

Building Block—Process, People, and Technology

The three components—processes, people, and technology define the foundation of an organization's CAB business strategy, Fig. 8.2. The component people comprises of leadership, talent management, and culture; process deals with CAB ecosystem; technology covers the aspects of cloud, analytics [models], and big data infrastructure (virtual, physical)

Leadership has to transform the culture toward embracing the CAB framework. If a leader or decision maker takes a back seat, then middle or junior manager will have less chance of initiating any efforts in that direction. Leaders are responsible for developing business strategy, articulate compelling and competitive vision, and

Fig. 8.1 CAB framework execution elements

Fig. 8.2 CAB business
strategy components

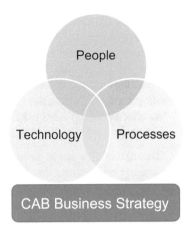

build a culture to transform and ensure that organization is moving in a right
direction (Gehani, 2011; Latham, 2012). The UK retailer Tesco has set up a robust
data-driven culture across the organization having coverage from its top leadership
to frontline. The CAB leaders, executives, and champions have distinct characteristics to drive the CAB adoption and leadership.

- Passionate about CAB ecosystem: The leaders are passionate to work with CAB
  ecosystem, and they understand the value of it and have know-how about the
  CAB power to excel the business value (Fig. 8.3).
- Drivers of the CAB ecosystem: The leaders are the drivers of the CAB
  ecosystem. They are aware of the business use case support and can align
  strategic and tactical decisions to power the [CAB] business strategy.
- Executors of the CAB outcome: They are capable of understanding and executing the CAB decision outcomes. Differentiation is achieved not when you

Fig. 8.3 Business value
matrix *BUC = Business Use
Case

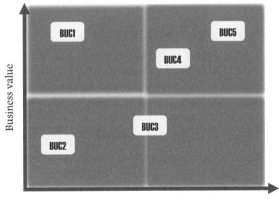

identify there exists a different customer segment but when you provide different offerings and value-added service to the different segments. The leaders have intrinsic interest and ability to utilize outcome in a more meaningful way.

- Edge at CAB space: The leaders are always abreast with the new trends and technology related to the CAB ecosystem. As more and more capability and competency pooled in the CAB space of the organization, these leaders do not get affected by the other involvement preferably they are prepared to excel in the way to lead the CAB ecosystem.

The organization needs to retain, recruit, and train people to build the required competency and skillset for the CAB adoption. It is better to identify the gaps regarding the necessary resource and competency before planning for the talent recruitment life cycle. The dynamics of the CAB space is too complicated, and thus it is difficult for the organization to retain and manage the right competency (Choi & Bae, 2009). A CAB knowledge-sharing collaborative system needs to be deployed where people from all level can discus about their experiences, findings, challenges of working in the CAB space and seek mentorship, guidance, and support to perform and excel.

More importantly, organization culture that contributes to the social and physiological environment impacts the transition and transformation realization outcomes. Until all people feel inclusive regarding the unified goal, business value, benefits, and competitiveness, the organization will not be able to materialize its core CAB values fully and attain leadership. The leaders should set an example by demonstrating the value of CAB. For example, Harrah's changed its culture by incorporating data and facts based decisions in rewarding employees which were previously run by paternalism and tenure. The organization has to put the effort into enterprise-wide CAB adoption and availability of centralized resources to get the maximum benefit. Walmart has several analytics groups working on different business-related problems, but now emphasizing in coordinating and centralizing analytics activates. Sears Holding Corporation has established centralized analytics team to work on analytics-driven projects.

To improve the organization culture, leaders should also focus on its partners, collaborators, vendors, and complete supply chain involvement to align them with its [CAB] business strategy. An organization having short of IT competency tends to outsource most of the IT-driven activities. For example, the organization may choose to the source of the public cloud to execute and deploy analytical models. Some organization act as partners and provide support to establish, manage, and execute CAB ecosystem. For example, SAS and Teradata help organizations to implement their solutions offerings. On the path of CAB adoption, organization has to develop an action plan regarding:

- Identifying core capabilities and competencies
- Developing capabilities and competencies
- Outsourcing capabilities, activities, and workflows
- Finding core partners dealing with CAB solutions

- Creating accessibility and accountability norms
- Preparing short-term, middle-term, and long-term security and control mechanisms, policies, contracts, and alliance
- Prioritize business use cases

Business users must prioritize the use case based on its business value and system and implementation complexity. It will help to strategize the roadmap and actions plans for handling the business use cases. Also, the stakeholders will get aware of the primary, secondary, and other use cases.

CAB Maturity Model

All sorts of domains and sectors are getting impacted by cloud, analytics, and big data. No matter whether you are in the business of CAB offerings or solutions sooner or later you have to enter into this market. You need to be prepared not just for the entry into the competing market but be ready to leap. The CAB maturity model (CABMM) will help you and your organization to understand status-quo, and by utilizing it organization can move incrementally or leap-frog to attain significant competitiveness. The CABMM is described through seven different levels of CAB maturity and cut through different elements of a business ecosystem.

Seven Levels of CAB Maturity

The management structure that is composed of different levels and specialties often has different levels of authority, responsibility, and stake. Management's responsibility is to understand the situation, make sense of it, identify problems, develop solution space, make action plans, and execute the strategic and tactical decision. There are seven different CAB maturity levels (Fig. 8.4):

- Level 1: Ad hoc: At this level, organization culture is risk-averse, and leaders are not keen on foraying into the CAB space. Leadership vision is short and mostly driven by the past performance goals and settings. They are reluctant to take any initiatives. People are not encouraged to develop concepts and execute POCs. There is no sign of using CAB-related tools either for demo's purpose or trial run. At this level, leaders do not think of providing CAB-related training, and they consider it not to be a fruitful investment and thus do not provide any such environment. The main goal of leaders is to focus on current market share and not to innovate and improve. There is no mechanism and metrics related to the measurement of CAB awareness, sensitization, and operationalization capabilities
- Level 2: Staged: At this level, the organization's culture to appreciate CAB philosophy is present. Employees ideas related to CAB are heard and discussed

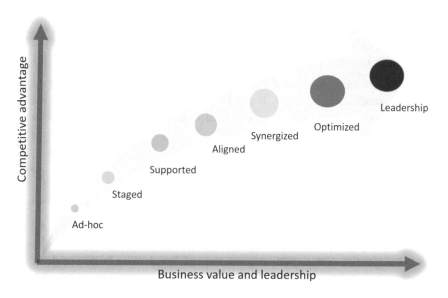

Fig. 8.4 CAB maturity model

at select few forums. Leaders are aware of the CAB trend but are hesitant to experiment as they do not have clear structure and competency in place. The organization does not have system and process to support people ideas to experiment with POCs and pilot project run. Sporadic initiation of CAB awareness training programmers is executed. Due to the lack of CAB business strategy, the ideas submitted are not aligned and thus discarded. Some advancements in the mechanism and metrics related to the measurement of CAB awareness are considered

- Level 3: Supported: At this level, the organization's culture is risk tolerant, and motivate and appreciate the people ideas related to CAB space. Leaders gain confidence in understating the benefits of the CAB adoption to attain the competitive advantage by providing CAB offerings to the customers, internal and external partners, collaborators, associates, and alliance. Everyone is motivated to discuss CAB inclusion to support and extend business value. Leader formulated special task force and steering committee to formally layout the concept note, draft roadmap of CAB adoption. Leaders take a more active role in setting the directions. Select few projects or business units are allowed to experiment or run POCs to verify and validate the feasibility of the CAB adoption. The select few CAB tools are available for the volunteers to perform test run and evaluate the benefits. Training is also planned at this stage to sensitize the people and at certain level build the competencies. For example, people will be trained in visualization, reporting and dash boarding activities, development of business use cases. Some advancements in the mechanism and metrics related to the measurement of CAB business use cases are considered.

- Level 4: Aligned: At this level, organization's culture is risk-taking and support collaborative knowledge-sharing system. People are motivated and appreciated in cross-functional involvement and experimentation of CAB-related activities, workflows. POCs, and pilot projects. The leader also has incorporated CAB elements in select few business units and has validated the success of business use case generating business values. They share the best practices and standards with other business leaders and business domains. Leaders themselves envision CAB deployment enterprise-wide and understand potential benefits. CAB business strategies, initiatives, business use cases are formulated. CAB steering committee and governance board are also established. Competency building (by training or recruitment) is the major exercise executed at this level. Leaders able to develop the business portfolio to be benefitted by the CAB adoption and have gained confidence in running few select business initiatives and use cases. Standard mechanisms and metrics related to the measurement of CAB business values and proposition have been formulated and utilized.
- Level 5 Synergized: At this level, organization's culture support collaboration, coordination, and communication environment. The organization understands to design, capture, and create the business value across its value chain and establish the value proposition. The processes and infrastructure required to support CAB initiatives are deployed. Multiple CAB projects across various business unit benefitting different stakeholders are executed. CAB workflows and processes are (semi)-automated and operationalized across significant business initiatives.
- Level 6 Optimized: At this level, organization's DNA is CAB driven. Every facet of the organization are touched and impacted by the elements of the CAB. The organization is on the transformation path and ready to foray into offering a CAB driven portfolio of services and products other than the core offerings. The organization enters a new market, penetrate new segment, and build new businesses.
- Level 7 Leadership: At this level, an organization set standards and benchmarks and considered as a leader in the CAB space in generating the business values. Fully automated and standardized process is deployed and executed. Executive leadership fosters innovation. The whole value chain and business ecosystem are supported and influenced by the organization CAB Portfolio.

Concluding Remarks

Your decision to adopt CAB framework requires carefully thought out transition and transformational planning. The leaders must envision the type of change and guide the organization to undergo such change. The mere adoption of CAB-related technologies does not help the organization to attain CAB leadership. It should be present as part of its DNA. It is paramount essential to the success of any business initiative that it must support the business strategy. The three components—processes, people, and technology—define the foundation of an organization's CAB business strategy. The CAB maturity model (CABMM) will help organization to understand status-quo, and by utilizing it organization can move incrementally or leap-frog to attain significant competitiveness. The CABMM is described through seven different levels of CAB maturity and cut through different elements of a business ecosystem.

References

Chen, C.-C., Shih, H.-S., Shyur, H.-J., & Wu, K.-S. (2012). A business strategy selection of green supply chain management via an analytic network process. *Computers & Mathematics with Applications, 64*(8), 2544–2557. https://doi.org/10.1016/j.camwa.2012.06.013.

Chen, Y.-K. (2012). Challenges and opportunities of internet of things. In *17th Asia and South Pacific Design Automation Conference* (pp. 383–388). https://doi.org/10.1109/ASPDAC.2012.6164978.

Choi, S. H., & Bae, S. M. (2009). Strategic information systems selection with incomplete preferences: a case of a Korean electronics company. *Journal of the Operational Research Society, 60*(2), 180–190. https://doi.org/10.1057/palgrave.jors.2602537.

Gehani, R. R. (2011). Management & innovation individual creativity and the influence of mindful leaders on enterprise innovation, *6*(3).

Gupta, P., & Consulting, A. (2010). Business innovation maturity model (BIMM), 2010.

Kshetri, N. (2014). The emerging role of Big Data in key development issues: Opportunities, challenges, and concerns. *Big Data & Society, 1*(2), 2053951714564227. https://doi.org/10.1177/2053951714564227.

Latham, J. R. (2012). Management system design for sustainable excellence: Framework, practices and considerations. *Quality Management Journal, 19*(2), 7–21. Retrieved from http://search.ebscohost.com/login.aspx?direct=true&db=bth&AN=75281631.

Upadhyay, N., & Deshpande, B. M. (2010). SDCS: Six-dimensional classification strategy framework for COTS products. *International Journal of Data Analysis Techniques and Strategies, 2*(2), 170. https://doi.org/10.1504/IJDATS.2010.032456.

Chapter 9
CAB Control—The Power

The Stone Age didn't end because they ran out of stones....
—Sheikh Yamani

All over the place, from the popular culture to the propaganda system, there is constant pressure to make people feel that they are helpless, that the only role they can have is to ratify decisions and to consume.

—Noam Chomsky

Introduction

Organizations face a cultural shift when dealing with CAB framework. The CAB control mechanisms can shape the corporate philosophy in terms the way cloud, analytics, and big data to be utilized to drive the business strategy vision and business values (Löbler, Gomes, Pozzobon, & Gomes, 2012; Mariano, Junior, Biancolino, & Maccari, 2013). Both the technology and business users have to collaborate to define the CAB control mechanisms, more specifically CAB governance to identify the elements and rules to interact with the elements. The potential benefits of CAB framework are compelling. The company target initiated high-performance marketing by incorporating customer data with other available data sources by running analytical processes to better segment the customers and to precisely target the recommendations. It was discovered that during pregnancy, customers likely to form different shopping habits. Most of the time customers were approached, for offers, promotion activity, and discounts only after when a child was born and thus it was too late to capture and delight the customer. Target developed a pregnancy prediction model that could predict the likelihood of the pregnancy based on the "shopping basket." It ran the model on the its customers list and after finding good performance started using it for the potential customers by sending promotional offers for the baby items. By doing such kind of marketing activity, target didn't violate any law but it didn't take into account the age of the customer. In one case, a father furiously reacted when his daughter, who was a teenage girl, received promotional offers for the baby items by the target.

© Springer Nature Singapore Pte Ltd. 2018
N. Upadhyay, *CABology: Value of Cloud, Analytics and Big Data Trio Wave*,
https://doi.org/10.1007/978-981-10-8675-5_9

Fig. 9.1 CAB controls—
governance

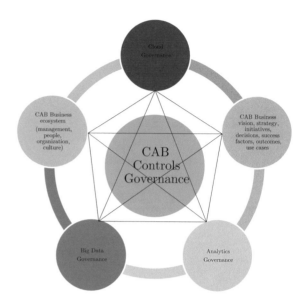

Nonetheless, target was correct as later father confirmed about his daughter's pregnancy. In essence, target miss the appropriate controls to utilize the [big] data and analytical models for its marketing campaigns and received heat from the customers.

CAB is not just mere a technical competency but more a business competency and the business users are the key stakeholders of the CAB adoption framework.

CAB Governance

CAB control mechanisms are important and are driven by the CAB governance framework Fig. 9.1. It has five elements:

- Cloud Governance
- Analytics Governance
- Big Data Governance
- CAB Business ecosystem
- CAB vision

Cloud Governance

Cloud governance is vital to the success of your cloud adoption or sourcing strategy. To avoid fragmented implementation having minimal impact and to achieve the maximum benefit by the cloud adoption or sourcing, you need to develop the business value plan by considering the external forces, current infrastructure, resources competency and capability and culture (Hashem et al., 2015; Nam & Pardo, 2011; Zwitter, 2014). Few examples of why one should achieve cloud governance are:

- Determining the necessity and value decision of cloud adoption
- Identifying the ownership of cloud offerings
- Ensuring the execution of cloud service ownership responsibilities
- Achieving services, applications, and processes migration
- Enforcing accountability and policies for the cloud engagement
- Establishing SLAs
- Selecting decision makers for cloud adoption
- Deploying functional and cross-functional cloud adoption strategy

Cloud landscape also often involves outsourcing, thus organization must pay attention to the outsourcing opportunity and challenges. Two-point agenda is critical for the business leaders:

- Identification of critical success factors impacting the business
- Examining ways to govern the cloud

The adoption of cloud should not be done without gaining the confidence of IT and business leaders, otherwise it will be very challenging for the organization to maintain the harmony between the two groups. Most of the time organization keeps cloud management as a part of cloud governance. Cloud governance deals with the enablement of the business environment by setting directions to use cloud for the business functions and business value. Cloud management refers to the execution of the cloud governance directions. Cloud service needs to be treated differently than the traditional IT service management and governance (Jula, Sundararajan, & Othman, 2014; Sharma, 2015; Vouk, 2008). Table 9.1 shows the impact of cloud services on the governance and management.

The organization will also have to redefine the roles and responsibilities as per the cloud service model adoption or sourcing, Table 9.2.

It is better to set up a special leadership task force for cloud governance and implementation (CGI) to avoid any conflicts among the stakeholders. It is often due to the absence of special task force, like CGI, the outcomes are inadequate directions and planning, unclear actions, and oversight resulting in failure of IT initiatives. Following points help organization to charter the CGI and directions for cloud adoption or sourcing:

- Appointing CGI leader and planning a charter: A CGI team leader is appointed who can either be from IT or business unit within the organization but keen and motivated to adopt or source the cloud service. A CGI leader needs to collaborate and communicate across all the levels of the organization and responsible for recruitment and training of the team members. CGI team is also responsible

Table 9.1 Cloud service impacts on governance and management

Cloud service characteristics	Impact on traditional ITSM and governance
On-demand self-service	• New ways of service monitoring, tacking, and management • Deployment of security, privacy, and policy controls
Board network access	• Formulation of stringent operational-level agreements • Active communication and engagement for support, configuration, and management activities • Designing and devising of policies, contracts, and processes
Resource pooling	• Consideration of multitenant environment • Business processes, services and applications migration decision
Rapid elasticity	• Identification of demand management and cost budgeting • Strategic usage of compute and elastic resource
Measured service	• Integration of metered service and chargebacks • Deployment of SLAs
Service model	
Cloud service	• SaaS model provides wide scope of service to the cloud customer as compared to the IaaS • PaaS model implementation has impacts to business processes and administration

Table 9.2 Cloud service models and responsibilities

Cloud Service	Responsibilities	
SaaS	Cloud customer	————
	Cloud provider	Manage entire stack of a cloud service such—Applications, data, runtime, middleware, operating system, virtualization, servers, storage, networking
PaaS	Cloud customer	Manage service layer such as applications, data
	Cloud provider	Manage development and infrastructure platforms such as—runtime, middleware, operating system, virtualization, servers, storage, networking
IaaS	Cloud customer	Manage service and development layer such as—applications, data, runtime, middleware, operating system, virtualization
	Cloud provider	Manage infrastructure platforms such as—servers, storage, networking

to charter the growth and implementation charter for the cloud adoption or sourcing.

• Formulating CGI implementation team: The involvement of business and IT unit representatives is required for understanding business requirements and challenges. The CGI team should be comprised of at least—executives, managers, and practitioners. Executives are the key stakeholders having direct interest and stake in the cloud adoption or souring. They are also responsible to review and act on the progress of the cloud adoption. Managers look at various tasks related to cloud service planning, service building and migration, service deployment, and service monitoring. Practitioners are the backbone of CGI implementation team and drive the cloud adoption or sourcing operational processes.

Organizations play varied role during the cloud adoption or sourcing business strategy life cycle. The roles can be different at pre-adoption, adoption, and post-adoption phase.

- Cloud service provider: It is responsible to provide cloud services to internal or external customers.
- Cloud service broker: It is responsible to provide value-based cloud services to internal or external customers.
- Cloud service integrator: It deals with aggregation of the data and data services.
- Cloud service consumer: It is the one that consume the cloud services based on the requirements.
- Cloud service regulator: It is responsible to regulate the cloud governing environment.

All organizations are not alike and thus with varied need each organization has to define its own set of principles for the effective governance. A principle serves as a foundation to establish decision making policies which further elaborates the rules under which an organization has to work during cloud adoption and usage, Table 9.3.

Analytics Governance

The analytics function, capability, and competency in many organizations are rapidly growing. One must address the challenges and concerns unique to the analytics. Many organizations face challenges in maintaining a pipeline of analytics projects considering multiple data sources (Dutta & Bose, 2015; Iqbal, Doctor, More, Mahmud, & Yousuf, 2016; Niederer & Chabot, 2015; Singh & Singh, 2012). Analytics have several stakeholders—executives, managers, analytics practitioners, data management practitioners, end users, customers, almost every one of the organization (associated internally and externally) gets affected by the analytics, Table 9.4.

The stakeholders need to coordinate, collaborate, and communicate to align their interests, and expectations at the beginning of the analytics journey. It will help them to understand any risks associated with operationalization of the analytics, select, and evaluate analytical models to be developed and utilized, streamline the projects timeline, prepare the right resources, identify the scope and constraints, and generate value proposition. A five phase of analytics governance, Fig. 9.2, once adopted and executed will provide to organization an edge, competitiveness, and operationalization of analytics. At each phase, the business leaders and associated stakeholders can share intermediate feedback, learnings, and know-hows.

Table 9.3 Example cloud service guiding principles

Principle	Description
Alignment	To align business vision, strategy, initiatives, use cases, critical success factors with the cloud service
Communication and reporting	To develop platform to communicate and report to the concerned stakeholders seamlessly
Acquisition	To ensure cloud service and business needs are met by acquiring cloud services
Agility	To fully leverage cloud service capability and functionality
Risks	To ensure the risks related to cloud service, for example, security, privacy, safety, data, and information sensitivity are identified, tracked, controlled, and mitigated
Compliance	To ensure cloud service utilization and actives are as per ethical, legal, and regulatory purview
Engagement	To ensure effective engagement of stakeholders
Training	To ensure teams and resources are trained and have competency to drive cloud-related projects
Performance	To track, monitor, execute, and enforce operational-level agreements and service-level agreements
Authority	To ensure that authorized stakeholders are engaged with cloud projects, cloud service usage and management, cloud service coordination and control
Accountability	To ensure that assigned responsibilities are executed and met under the purview of control and monitoring mechanisms

Table 9.4 Stakeholder's analytics challenges and concerns

Stakeholder	Challenges and concerns
Manager	How to operationalize analytics effectively? How to solve business issues? How to ensure to achieve the analytics outcome? How to manage pipeline of analytics projects?
Analytics practitioners	How to streamline analytics format and structure outcome? How to coordinate multiple and disparate data sources? How to manage pipeline of data sources to multiple business units and multiple analytics projects?
Data management practitioners	How to address analytics ad hoc request from managers and analytics practitioners? How to manage, control, and utilize data infrastructure and resources? How to perform and act on data audits?
Business leaders	How data and information be used? How privacy, safety, security, and ethical concerns are addressed? How to envision business analytics leadership? How to establish analytic-driven culture?

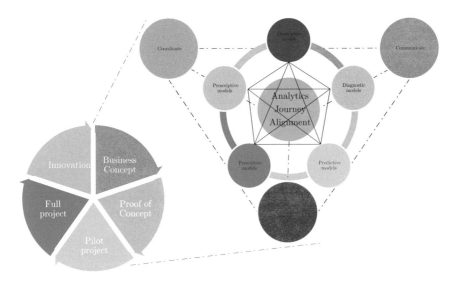

Fig. 9.2 Analytics journey alignment

- Business concept: At this phase, business leaders need to define the analytics problem, issue, and requirement; establish analytics strategy (project or enterprise wide; incremental or big bang, etc.); plan analytic tasks, role, and responsibilities.
- Proof of concept (PoC): In this phase, develop and execute controls, access and management mechanisms; analyze data and build multiple models; select and evaluate performance of models; identify new business processes affected by the models; assess and quantify business impact and value.
- Pilot project: At this phase, rollout the project execution; communicate to stakeholders; identify and define KPIs, KAGIs, tune the performance parameters of the models.
- Full project: In this phase, deploy and monitor the full project; check and tune the performance parameters; execute change management; design new business processes; achieve semi/full automation.
- Innovation: At this phase, change and update analytical models based on trends, business requirements, competitiveness, new business processes, and operationalization scope and constraints.

Each stakeholder will benefit by the analytics alignment. At this stage, key analytics governance initiatives (KAGI) covering value proposition are to be formulated which needs to be linked and mapped with the analytics projects and organizations capabilities and competencies, Table 9.5.

Firstly, a roadmap to analytics journey to be created. It should be any time between nine months to five years depending upon the vision of the organization. Further, the roadmap must include the type, outcomes, cost, scope, and constraints of analytics projects. The analytics projects fall into five categories—descriptive,

Table 9.5 Stakeholder's benefits of analytic alignment

Stakeholder	Benefits of analytic alignment
Manager	Prioritize analytics project pipeline Develop and streamline effective timeline Identify scope and constraints Charter and prepare outcomes Plan and control costs, and resources
Analytics practitioners	Establish clear guidelines of expected outcomes and data availability Prepare and develop control points to manage, control, and utilize data infrastructure and resources Achieve seamless collaboration with managers and data management practitioners
Data management practitioners	Prepare, develop, select, and execute analytical models Ensure delivery of the data and information as per the scheduled agreed Plan to manage, control, and utilize data infrastructure and resources? Develop action plan to perform and act on data audits
Business leaders	Develop guidelines for the usage of data and information Develop privacy, safety, security, and ethical control mechanisms and action plans Envision business analytics leadership by collaborating, communicating, and coordinating vision and strategy Establish analytic-driven culture by operationalizing analytics and by building analytics competency

diagnostic, predictive, preventive, and prescriptive models (Chae, 2015; Crawford & Schultz, 2014; Wang, Gunasekaran, Ngai, & Papadopoulos, 2016). Each project needs to be aligned with the business vision, strategy, initiatives, decisions, success factors, and use cases. Identifying and prioritizing the analytics projects based on costs, implementation feasibility, system complexity, and resource constraints avoid any conflict and challenges in executing the analytical projects.

Figure 9.3 shows sample description of the analytics roadmap.

Secondly, the analytics roadmap must be synchronized alongside with the timelines, data and infrastructures requirement, and development of resource competency. Lastly, the governing board is required to control and govern the activities charted in the analytics roadmap. The board is required to assess each analytic project business impacts, associated costs and resources, timeline and operationalization feasibility.

Big Data Governance

The governance for big data is required to fulfill the needs for corporate governance, IT governance, cloud, and analytics governance. In absence of the big data governance (BDG), the outcome of the CAB adoption or sourcing strategy will not produce [effective] business values to businesses. The BDG requires a different set of guidelines, principles, and criteria to drive big data projects. The focus will be on

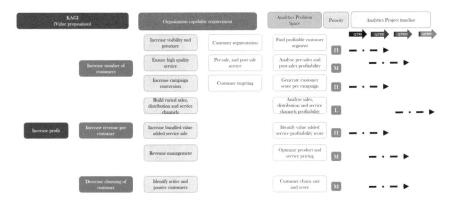

Fig. 9.3 Analytics roadmap

big data services, quality of big data, security, privacy, and accountability of sensitive and disclosed data. According to McKinsey, more specifically to the case of USA, a further improvement of productivity by 1% is estimated by the utilization of big data in health care, public administration, retail, and manufacturing and personal information sectors contributing to economic scale of worth at least $100 billion to $700 billion. The purpose of BDG is to establish infrastructure to support big data services and provide stable quality services through a balance of responsibility and authority. Business data leaders must consider big data attributes while dealing with big data.

- Timeliness: It deals with the timely collection, sourcing, aggregation, and wrangling of the data for the analysis purpose.
- Trustfulness: It refers to the reliability of the data to provide inferences and business insights.
- Meaningfulness: It deals with the usage of data to provide meaningful information and actionable insights.
- Sufficiency: It refers to the availability of space and time resource to provide analysis, inferences, and actionable insights.
- Accountability: It provide details about who does what, when, why, and how with the data at rest, on mode and during process.
- Accessibility: It ensure that data is accurate, available, timely, and of direct value to stakeholders.

The BDG team must be responsible to provide nomenclatures, terminology, definitions, policies to control, manage, access and distribute [big] data, and perform routine review of governance for enhancements. The BDG must also in the alignment of cloud and analytics governance. There are four elements of BDG framework, Fig. 9.4:

- Big data governance board: It refers to the team responsible to charter the initiatives, implementation, controls and security mechanisms, growth plans and ensure competency to deal with the big data projects.

Fig. 9.4 Big data governance

- Big data governance vision: It deals with the business vision of the big data to generate business value. The introduction of the big data technology without aligning it with the big data vision, policies, and organizational values can increase risks.
- Big data governance strategy: It helps to align big data initiatives, outcomes, success factors, decisions, and use cases with the business strategy.
- Big data components: It comprises of management, principles and policies, and standardized processes. Further, the components are responsible to define data formats, storage and processing policies, supervision and management of data flow, control and execution of data flow and processes, quality and accuracy of data, authority, responsibility and accountability of the data.
- Big data infrastructure: It deals with the twofold structure. Firstly, it is driven by the big data audit team that verifies and validates the big data projects as per the compliance, legal, and regulatory requirements. Secondly, the infrastructure required to collect, store, process, analyze, and visualize data which is managed by on premise, on cloud, or on hybrid infrastructure.

Concluding Remarks

The CAB control mechanisms can shape the corporate philosophy in terms the way cloud, analytics, and big data to be utilized to drive the business strategy vision and business values. Both the technology and business users have to collaborate to define the CAB control mechanisms, more specifically CAB governance to identify the elements and rules to interact with the elements. The potential benefits of CAB framework are compelling. CAB is not just mere a technical competency but more a business competency and the business users are the key stakeholders of the CAB adoption framework.

References

Chae, B. (2015). Insights from hashtag#supplychain and twitter analytics: Considering twitter and twitter data for supply chain practice and research. *International Journal of Production Economics, 165,* 247–259.

Crawford, K., & Schultz, J. (2014). Big data and due process—Toward a framework to redress predictive privacy harms. *BCL Rev., 55,* 93–128. https://doi.org/10.1525/sp.2007.54.1.23.

Dutta, D., & Bose, I. (2015). Managing a big data project: The case of ramco cements limited. *International Journal of Production Economics, 165,* 293–306. https://doi.org/10.1016/j.ijpe. 2014.12.032.

Hashem, I. A. T., Yaqoob, I., Anuar, N. B., Mokhtar, S., Gani, A., & Ullah Khan, S. (2015). The rise of "big data" on cloud computing: Review and open research issues. *Information Systems, 47,* 98–115. https://doi.org/10.1016/j.is.2014.07.006.

Iqbal, R., Doctor, F., More, B., Mahmud, S., & Yousuf, U. (2016). Big data analytics: Computational intelligence techniques and application areas. *International Journal of Information Management.* https://doi.org/10.1016/j.ijinfomgt.2016.05.020.

Jula, A., Sundararajan, E., & Othman, Z. (2014). Cloud computing service composition: A systematic literature review. *Expert Systems with Applications, 41*(8), 3809–3824. https://doi. org/10.1016/j.eswa.2013.12.017.

Löbler, M. L., Gomes, B., Pozzobon, D. M., & Gomes, C. M. (2012). Strategic orientation towards sustainable innovation: A case study in a brazilian university. *Journal of Technology Management & Innovation, 7*(2), 196–206.

Mariano, A., Junior, C., Biancolino, C. A., & Maccari, E. A. (2013). *Cloud Computing and Information Technology Strategy, 8,* 178–188.

Nam, T., & Pardo, T. A. (2011). *Conceptualizing Smart City with Dimensions of Technology, People, and Institutions* (pp. 282–291). https://doi.org/10.1145/2037556.2037602.

Niederer, S., & Chabot, R. T. (2015). Deconstructing the cloud: Responses to big data phenomena from social sciences, humanities and the arts. *Big Data & Society, 2*(2), 2053951715594635. https://doi.org/10.1177/2053951715594635.

Sharma, S. (2015). Evolution of as-a-Service era in cloud, (Figure 3), 1–14.

Singh, S., & Singh, N. (2012). Big data analytics. In *2012 International Conference on Communication, Information {&} Computing Technology (ICCICT)* (pp. 1–4). https://doi.org/ 10.1109/ICCICT.2012.6398180.

Vouk, M. A. (2008). Cloud computing—Issues, research and implementations. In *Proceedings of the International Conference on Information Technology Interfaces*, ITI (pp. 31–40). https:// doi.org/10.1109/ITI.2008.4588381.

Wang, G., Gunasekaran, A., Ngai, E. W. T., & Papadopoulos, T. (2016). Big data analytics in logistics and supply chain management: Certain investigations for research and applications. *International Journal of Production Economics, 176,* 98–110. https://doi.org/10.1016/j.ijpe. 2016.03.014.

Zwitter, A. (2014). Big data ethics. *Big Data & Society, 1*(2), 2053951714559253. https://doi.org/ 10.1177/2053951714559253.

Chapter 10
CAB Success Stories

It had long since come to my attention that people of accomplishment rarely sat back and let things happen to them. They went out and happened to things.

—Leonardo da Vinci

The dynamic nature of knowledge is reflected in human progress and technological achievements.

—Eraldo Banovac

Airbnb

Source: https://aws.amazon.com/solutions/case-studies/airbnb/

https://www.airbnb.co.in/

Airbnb has the motive to reinvent the travel and community-driven market place. By bringing host and guest together at a common platform, the customers and travelers are achieving high and unique level of personalization and local experience. The Airbnb was started in the year 2008 by Brian Chesky, Joe Gebbia, and Nate Blecharczyk, and it took almost 4 years for Airbnb to reach to the guest service of customer base of around 4 million. It hosts around 150000 people on its platform around the world to provide them a seamless experience of finding the right places and rights hosts/customers. Aribnb has more than 5 million listings across worldwide covering 81000 cities in more than 191 countries with all-time guest arrivals of crossing 300 million. Initially started as Airbedandbreakfast.com they officially appeared during SXSW and made two bookings. In 2009, Airbnb got a name changed from Airbed and breakfast. The first move to penetrate the digital market happened when it launched its first iPhone application in 2010. To attract the customers, Airbnb followed customer-centric strategy and introduced the $1 million host guarantee which helped many hosts to list their property at Airbnb. It also launches the wish list feature to track what customers and travelers are

N. Upadhyay, *CABology: Value of Cloud, Analytics and Big Data Trio Wave*, https://doi.org/10.1007/978-981-10-8675-5_10

interested in as part of their traveling experience considering homestay. In 2014, Airbnb got a big hit when it hosted more than 100,000 guests during the Rio World cup and became an official alternative accommodation supplier for 2016 Rio Olympic games.

The Airbnb faced huge challenges in the infrastructure support as their customer base and property listings kept growing. One of the specific challenges mentioned by Mike Curtis, VP of Engineering Airbnb, is to manage the "infrastructure scaling." The challenge was to scale the infrastructure to support the growing demand of the customers and to partner with the trustworthy, reliable, and credible cloud solution provider company. They have adopted Amazon Web Service to provide their travel solutions to the customer and hosts. Amazon provides a portfolio of solutions to the numerous clients across the world. Some of the solutions that Airbnb adopted are:

- EC2 Servers, production, Traffic, Hive Hadoop (2010: 24 EC2 instances; now more than 1000 and growing)
- RDS: All relational databases (2 billion rows in RDS)
- S3 Storage UGC: Photo stored in S3 (business case: High-quality photo vs low-quality has impact on the property listings and selections; 2010 300 GB and now more than 50 TB
- Dynamo DB Rollups and metadata
- Elastic Cache MemCache

The number of EC2 instances to offer the services to the customers increased drastically from 24 in the year 2010 to more than 1000 instances now. They identified that the high-quality property photo put up a promising profitable opportunity as compared to the low-quality photo. Such a proposition demanded to have a huge capacity of the infrastructure to store high-quality photo. Their demand to store photo grew from 300 GB in the year 2010 to 50 TB now. Big data and analytics were used across the company to consider decisions. Airbnb analytical model could understand host's preference for the accommodation request. The model provides the likelihood of the host's accommodating a customer based on the customers search engine query.

Mike Curtis, VP of Engineering Airbnb, explained that due to the promising infrastructure support from Amazon, it could invest all its efforts, resources, and time in building the unmatchable experience to the hospitability industry.

> People will think and feel that they have home everywhere where ever they want to travel in the world

Airbnb uniquely adopted the cloud, analytics, and big data technology ecosystem to offer the seamless access to the property listings and to empower millions of people by providing an opportunity to become hospitality entrepreneurs.

Learnings

Prompt and proactive
Leveraging infrastructure at scale
Achieving operational excellence
Gaining strategic advantage

Clearsense

Source: https://clearsense.com/news/healthcare-analytics-firm-gears-up-private-cloud-in-teirpoint-data-center/

Clearsense believes in providing real-time predictive analytics by engaging customers through its service offerings of "Data Ingestion Service (DIS)" and "Insight Ecosystem." The solution service provides an opportunity to healthcare system professionals to optimize their services and operations for a better health care and to make better clinical decisions. The DIS provides the facility to bring disparate system's data, aggregate, integrate, and normalize it to be utilized for the business use case. In addition, the IE service handles the veracity of the data so that always the newest version of the data set is being considered for the predictive analytics.

The critical challenge of the healthcare industry is to get the rich, timely, and correct data that could be used by the health care professionals to provide optimal services in the healthcare.

Clearsense is a company with a clear goal to help healthcare industry and professionals in providing better healthcare services. The company in its core offer smart and advanced analytics solutions.

The Clearsense partnered with the Hortonworks by utilizing its Connected Data Platforms—HortonWorks Data Platform and HortonWorks DataFlow(HDF) to provide real-time support of business use cases and streaming data to healthcare professionals. Inception is its flagship product that helps the key stakeholders of the healthcare industry to consider three critical areas:

- Clinical decisions: to generate informed and actionable decisions by providing timely information about the deteriorating conditions of the patients. It helped the professionals to achieve early detection of the problem related to patient healthcare and thus help to control it otherwise could lead to the catastrophic situations
- Financial decisions: to keep track of organizations KPIs and thus empowering professionals to take an informed decisions regarding financials
- Operational Decisions: to offer seamless and transparent processes across the care delivery system, thereby maintaining the high level of performance

The Clearsense developed cloud-based solution, Inception, which can be activated, accessed, and utilized from any technological environment. It helped Clearsense clients to get the benefit by using Inception's SMART data concept (Streaming, Measurable, Applied, Real-Time). Code blue was their critical cardiac care business use case application where the solution helped healthcare professionals to get aware of the potential cardiac related problem. For example, a cardiac caregiver could be notified about the status and health details of the person who is likely to get the cardiac arrest in 4–12 h.

Healthcare industry need to place a huge upfront cost in case they choose to opt for the proprietary license software. But with Clearsense solution, it was easy for the industry to use advanced analytics and provide better healthcare services as there was no upfront cost and the solution can be accessed on the real-time basis.

Learnings

Innovative and competitive advantage
Descriptive to predictive
Excellence in informed and actionable decision making
Gaining strategic advantage

Netflix

Source: https://www.netflix.com

(Teradata Corporation, 2014)

The digital entertainment company offers its service worldwide. It has more than 125 million subscribers in more than 190 countries. The viewers can watch more than 140 million hours of TV shows and movies per day, which includes original series, documentaries, featured films. The members have the luxury to utilize the Netflix entertainment service anytime, anywhere from almost on any platform and devices without any commercials. The viewer's data is collected, stored, and processed to understand the pattern, viewing habits, and likability of the next entertainment to be watched.

In 1997, Netflix was started as an Online Movie Rental service company and launched its Web site, Netflix.com, to offer DVD rental service. Deploying the subscription-based strategy, they introduced unlimited rental of DVD for a monthly charge. In 2000, Netflix introduced movie recommendation system that helped the company to understand Netflix member's choice based on its members movie rating.

Netflix announced $1 million prize competition that helped the company to deploy the analytics model that could predict the viewers rating for the movie based upon their previous ratings. The model is under continuous modification and is the core part of its analytical engine of the recommendation system. Netflix also offered the feature of tagging the movie which benefited the company to get the notable elements categorizing the movie. Netflix could generate more than 80,000 parameters that redefined the genre and extended to capture the micro-genre.

Netflix partnered with Amazon and Internet Service Providers (ISPs) for the delivery of its entertainment content. The entertainment service was hosted through Amazon Web service and also mirrored across the world through other ISPs. The Netflix's content delivery network (CDN) is managed through AWS. Netflix had two choice of moving its data center service to the AWS. Firstly, by moving all the service "as-is" on the AWS, but this would entail the problem and limitations associated with the data centers to persist event after the migration. Secondly, by choosing the cloud-native approach, by rebuilding the services, technology and the way that they operate the services. Netflix adopted the continuous delivery model considering the microservice architecture. The Netflix culture support collaboration, cooperation, and leadership. Netflix decided to adopt cloud using the microleadership strategy, where each engineering teams was given the ownership to take the decisions using self-servicing tools. It helped the company to innovate and embrace the CAB enterprise-wide. The portfolio of data infrastructure that is hosted on AWS comprises of:

- Big Data Technologies: Hadoop, Hive, Pig, Teradata, MicroStrategy.
- Open-source applications and services: Lipstick and Genie
- Streaming and Computing: Spark

Further, Netflix architecture supports following components:

- Events data: Kafka
- Operations data: Casandra
- Fast storage: Teradata, druid, Amazon Redshift
- Amazon S3
- Data service: Kragle, Metacat
- Data processors: Hadoop, Hive, Pig, Spark, Presto
- Data Visualization: Tableau, R
- Data Exploration: R, Jupyter

Netflix's partnership with Amazon has paid of effectively as the company can concentrate on its core functionality, and with AWS it can scale the infrastructure as per the requirement. Netflix is recognized as a cloud-native and big data company and through such initiatives it is trying to materialize innovative business scenarios.

Learnings

Move from reactive to proactive
Attaining cloud-native
Continues delivery
Gaining strategic advantage

General Electric

Source: https://techcrunch.com/2015/09/29/ge-predicts-predix-platform-will-generate-6b-in-revenue-this-year/
https://sloanreview.mit.edu/case-study/ge-big-bet-on-data-and-analytics/
https://aws.amazon.com/solutions/case-studies/general-electric/
http://sites.tcs.com/big-data-study/ge-big-data-case-study/
https://sloanreview.mit.edu/article/avoiding-analytical-myopia/

The General Electric (GE) company is transforming its journey from the industrial manufacturing to the industrial Internet manufacturing with the motive of establishing as a digital company. It has forayed into the convergence of machines, data, and the Internet and started deploying sensors to gas turbines, generators, jet engines, and all its machines; connecting them to cloud and analyzing the flow of data (stream, edge, batch). The company believes that the cloud, analytics, and big data have a promising role into the company's ability to be unique and have differentiation factor within the oil and gas industry.

GE has a strong goal to migrate all its workloads related to its core abilities and processes. A 140-year-old company is rich in culture and is not hesitating to embrace the cloud, analytics, and big data enterprise-wide. It is planning to migrate more than 9000 workloads, including 300 disparate ERP systems. It aims to reduce its data center footprint by deploying the resources, workloads, and processes to Amazon Web services.

GE oil and gas division has achieved to reduce total cost of ownership by 52% just by migrating half of its core applications to the AWS.

Jim Fowler, CIO GE, in his keynote re:Invent mentioned that the company is converging its physical infrastructure with the logical and software infrastructure to provide the enhanced outcome and customer-centric services. Wind farm optimization is one of the projects that has been successfully deployed and promised effective utilization and optimization of wind turbines. GE gathers data about wind turbine, ambient temperature, pressure, weather, and local environment details to optimize the functioning of the wind turbines. Further, GE is building a new generation of machines by incorporating machine data with the engineering of the machine data. It helped GE to understand the design modification depending upon the usage of the device, its intended function, life cycle, and other parameters. GE is expecting to drive $15 billion revenue by 2020 from its software convergence with the physical infrastructure.

- Service management
- Network perimeter
- Risk-based security controls
- Self-service automation
- Financial transparency

GE started building in-house capability creating center for software excellence and by recruiting skilled people. Build vs Buy decision was prominent to use and deploy services and opted for AWS solutions for the hosting of its valuable services. GE wants to build machines that could be self-serviced, automated, secured, controlled, and offer unmatched user experience.

GE adopted strategy to offload most of the workloads and application on the cloud and remove the data center footprints. But would like to have few data center where GE believes that it can keep, manage, and control its secret and confidential data, processes, and services. GE industrial Internet of Things platform Predix offers value-added services by integrating big data, machine, and the Internet. It also provides Predix.io as Platform-as-a-Service so that customers and other associated can build their own customized application on top of the PaaS.

Learnings

Move from reactive to proactive
Shift from physical to digital
Transition from people to machines
Gaining strategic advantage

Appendix

Worksheet—I

The organizations entering into the CAB space must consider some guiding principles to ensure that their efforts and investments are aligned with the corporate priorities. Answer each of the questions with respect to your organization.

Q.1. What benchmark we would like to set by adopting the CAB framework.
Q.2. Which strategy will be our differentiator factor to promise business value in our product/service offerings.
Q.3. Do we have a unified platform accessible to all stakeholders? What is a constraint and limitation if any?

© Springer Nature Singapore Pte Ltd. 2018
N. Upadhyay, *CABology: Value of Cloud, Analytics and Big Data Trio Wave*,
https://doi.org/10.1007/978-981-10-8675-5

Worksheet—II

To manage the organization's transformation and growth approach, answer the following questions and establish the traceability of the relationships that generate the business value.

Q.1. Describe organization's vision, mission, goals and objectives.

Q.2. Describe organization's business functions and cross-functions

Q.3. Describe and list organization's key information and interaction value chain with partners, vendors, suppliers, customers and external entities

Q.4. Describe and list organization's tangible and intangible assets

Q.5. Describe and list organization's core business, information and process models

Q.6. Describe Organization's key performance, success and agility parameters

Q.7. Explain and Describe organization's resource – people, material, technology ontological structure.

Q.8. Describe business initiatives

Q.9. Describe critical success factors

Q.10. Describe and list business use cases

Q.11. Describe management drivers

Q.12. Describe technology [CAB] drivers

Worksheet—III

The key stakeholders must address the following questions before undertaking the CAB solution evaluation exercise.

Q.1. How do you select Cloud providers that are aligned with our business strategy?
Q.2. How do you develop, test and deploy analytical models promising to generate business value?
Q.3. How do you control, manage, and source big data and big data infrastructure services?

Q.4. How do you ensure stakeholders concern are addressed?

Q.5. What all data you have in place?

Q.6. What all you want to do with the data?

Q.7. What solutions can integrate data with decisions?

Q.8. How data is stored, managed and processed?

Q.9. Does the Big data solution (BDS) able to manage data with mobility and structure?

Q.10. How effectively is the BDS capable of processing in a distributed manner and achieve parallelization?

Q.11. How does the BDS integrate with legacy, and frontier technology solutions?

Q.12. Does a solution provider believe in ecosystem strategy?

Q.13. How a solution provider extend the partnership relationship in more meaningful ways?

Q.14. How effective BDS does support exploration, visualization, and consumption of the data sets?

Q.15. Does the solution intersect horizontally across business or also available for the vertical business unit's purpose?

Worksheet—IV

To mitigate any risks in attempting to establish a robust CAB ecosystem, an important step is to answer each of the questions with respect to your organization.

Q.1. How to manage operational processes?
Q.2. How to manage cross collaboration and joint responsibility?
Q.3. How to manage organizational change and transition?

Q.4. How to manage service level management (SLMs), service level agreements (SLAs), and integrated service management (ISMs)?

Q.5. How to manage risks?

Q.7. How to manage control, access, authority, and accountability?

Q.8. How to manage ownership?

Q.9. Develop the CAB classification policy to classify the levels of the:
analytical models to be deployed and executed

Q.10. Develop the CAB classification policy to classify the levels of the:
data and security, privacy and safety controls

Q.11. Develop the CAB classification policy to classify the levels of the:
Cloud services and deployment models

Worksheet—V

The organizations entering into the CAB space must consider some guiding principles to ensure that their efforts and investments are aligned with the corporate priorities. Answer each of the questions with respect to your organization.

Q.1. Describe the elements covered at the entry point, the organization and CAB solution provider should decide and negotiate on the legal and compliance terms and conditions
Q.2. Describe the elements covered at the exit point, the organization and CAB solution provider should decide and negotiate on the legal and compliance terms and conditions
Q.3. Describe the elements covered at the service point, the organization and CAB solution provider should decide and negotiate on the legal and compliance terms and conditions

Worksheet—VI

The organizations entering into the CAB space must consider some guiding principles to ensure that their efforts and investments are aligned with the corporate
priorities to monetize and generate business value. Answer each of the questions
with respect to your organization.

Q.1. What characteristics are important for the business use case?
Q.2. What are the priority of business use case in terms of first, second and higher order consequences?
Q.3. What big data can be used across business units?

Q.4. What category of stakeholder's get affected by the big data?

Q.5. How much compute and storage capacity is required?

Q.6. What are the core, critical, moderate and easy processes, workflows and applications required for the operational purpose?

Q.7. How much infrastructure load is witnessed during product and service accessibility? What is the peak time?

Q.8. What is the total cost of own ership of the IT infrastructure?

Worksheet—VII

The organizations entering into the CAB space must consider some guiding principles to ensure that their efforts and investments are aligned with the corporate priorities to understand customers, overcome barriers and gather data. Answer each of the questions with respect to your organization.

Q.1. Who are my customers and more importantly valuable customers?
Q.2. What channels my customer use to utilize my services and products?
Q.3. What products and services are contributing to my customer's satisfaction?

Q.4. What are the barriers to the adoption of technology?

Q.5. What are the challenges in the offerings of products and services?

Q.7. The organization must identify the [big] data that it has or can get to utilize it to drive the business use cases.

Operational data

Enterprise "Dark" data

Public data

Commercial data

Social media data

Worksheet—VIII

The organizations entering into the CAB space must consider some guiding principles to ensure that their efforts and investments are aligned with the corporate priorities to bring culture change and transition. Answer each of the questions with respect to your organization.

Q.1. What marketplace are you in?
Q.2. What marketplace are competitive?
Q.3. What differentiators your products have?

Q.4. How do you improve and compete in the marketplace?

Q.5. What opportunity space is available?

Q.6. What are the challenges to foray into the marketplace?

Q.7. Identifying core capabilities and competencies

Q.8. Identify required capabilities and competencies for CAB adoption or sourcing

Q.9. Find core partners dealing with CAB solutions aligned with appropriate culture

Q.10. Creating accessibility and accountability norms

Q.11. Preparing short-term, middle-term and long-term security and control mechanisms, policies, contracts, and alliance.

Q.12. Prioritize business use cases

Index

Printed in the United States
By Bookmasters